POST-HARVEST TECHNOLOGIES
OF FRUITS & VEGETABLES

HOW TO ORDER THIS BOOK

BY PHONE: 877-500-4337 or 717-290-1660, 9AM–5PM Eastern Time

BY FAX: 717-509-6100

BY MAIL: Order Department
DEStech Publications, Inc.
439 North Duke Street
Lancaster, PA 17602, U.S.A.

BY CREDIT CARD: American Express, VISA, MasterCard, Discover

BY WWW SITE: http://www.destechpub.com

Post-harvest Technologies of Fruits & Vegetables

HOSAHALLI S. RAMASWAMY, PHD, FCSBE, FAFST (I)

Professor
Department of Food Science and Agricultural Chemistry
McGill University

DEStech Publications, Inc.

Post-harvest Technologies of Fruits & Vegetables

DEStech Publications, Inc.
439 North Duke Street
Lancaster, Pennsylvania 17602 U.S.A.

Copyright © 2015 by DEStech Publications, Inc.
All rights reserved

No part of this publication may be reproduced, stored in a retrieval system, or transmitted, in any form or by any means, electronic, mechanical, photocopying, recording, or otherwise, without the prior written permission of the publisher.

Printed in the United States of America
10 9 8 7 6 5 4 3 2 1

Main entry under title:
 Post-harvest Technologies of Fruits & Vegetables

A DEStech Publications book
Bibliography: p.
Includes index p. 311

Library of Congress Control Number: 2014940614
ISBN No. 978-1-932078-27-5

Contents

Preface ix

Acknowledgements xi

1. An Overview of Postharvest Losses and Causes 1
   Introduction   1
   Causes of Postharvest Losses   7

2. Classification of Fruits and Vegetables 11
   Definitions   11
   Classification of Fruits and Vegetables   13

3. Structure of Fruits and Vegetables 23
   Introduction   23
   Structure of a Plant Cell   23
   Structure of Fruits and Vegetables   26
   Texture of Fruits and Vegetables   31
   Texture-Structure Relationships   35

4. Composition and Nutritional Qualities 37
   Introduction   37
   Water   39
   Carbohydrates   40
   Proteins   46
   Fats and Oils   46

Organic Acids   50
Vitamins and Minerals   51
Pigments   53
Flavor Compounds   56
Anti-Nutritional Factors in Foods   57

5. Harvesting of Fruits and Vegetables 59
Introduction   59
Hand Harvesting   59
Mechanized Harvesting   60
Elements of Harvesting: Hand/Mechanical   62
Important Field/Orchard-Considerations for Mechanized Harvest   62
Types of Mechanical Harvesters   64
Harvest Aids   64
Extent of Mechanization   65

6. Quality and Maturity Indices 71
Introduction   71
Factors Influencing Quality   75

7. Post Harvest Physiology of Fruits and Vegetables: Respiration...................... 85
Introduction   85
Aerobic vs Anaerobic Respiration   86
Energy Currency—ATP   86
Aerobic Oxidation of Glucose   87
Biochemical Pathways   89
Anaerobic Respiration   93
Factors Influencing Respiration   94
Control Measures to Minimize Respiratory Losses   104
Typical Calculations Involving Respiration Problems   104

8. Post Harvest Physiology of Fruits and Vegetables: Transpiration 107
Introduction   107
Degree of Cooling Achieved by Transpiration   107
Significance of Transpiration During Plant Growth   108
Transpiration Represents Economic Loss   108

Transpiration Occurs through Specialized Tissues   109
Transpiration is a Diffusion Phenomenon   109
Partial Pressure, Saturated Water Vapor Pressure and
     Relative Humidity   110
Factors Influencing Transpiration   117
Methods of Controlling Transpiration Losses   119
Psychrometry   120

9. Cooling of Fruits and Vegetables … 127
Introduction   127
Modes of Heat Transfer   132
Cooling Rate   137
Cooling/Pre-cooling Methods   145
Principles of Refrigeration   154

10. Cold Storage Systems for Fruits and Vegetables … 167
Introduction   167
Refrigeration Requirements   177
Ventilated Storage   183
Refrigerated (Regular Atmosphere, RA) Storage   184
Control Atmosphere (CA) Storage Systems   189
Ethylene in Post-Harvest Technology   194
Membrane System for CA/MA   196
Hypobaric System for CA/MA   205

11. Packaging of Fruits and Vegetables … 211
Introduction   211
Package as a Handling Unit   212
Protection from Physical and Mechanical Injuries   214
Protection from Moisture Loss   217
Providing a Sanitary Environment   218
Facilitating Important Treatments   219
Prevention of Pilferage   219
Improvement of Sales Promotion   220
Communication with Consumers   220
Promotion of Product Identification and
     Company Recognition   220

Facilitation of Sales   221
Generation of Modified Atmosphere within the Package   221
Types of Containers   222
Types of Packaging Materials   223
Packaging Requirements for Fruits and Vegetables   230
Modified Atmosphere Packaging   231

12. Irradiation of Fruits and Vegetables 241
Introduction   241
Historical Perspectives   243
Forms of Radiation   243
Characteristics of Ionizing Radiations   248
Sources of Ionizing Radiations   250
Units of Radiation   255
Radiation Dose   258
Food Irradiation Worldwide   264
Detection of Radiation   273
Regulatory Approval   273
Economic Aspects   279
The Future   280

13. Postharvest Pathology 281
Introduction   281
Common Diseases   282
Disease Development   285
Mechanism of Pathogen Entry   287
The Infection Process   287
Defense Mechanism   287
Growth Behavior of Bacteria and Fungi   288
Treatments for Pathogen Control   290
Pre-Harvest Control   293

14. Postharvest Treatments 295
Insect Control   295
Chilling Injury   302
Freezing Injury   304

References and Reading Materials   307
Index   311

CHAPTER 1

An Overview of Postharvest Losses and Causes

INTRODUCTION

THE word "harvest" triggers many sensations among people in countries that depend on agricultural production. It is the beginning of realizing the gain from all the hard work that has been put in right from the time of planting, watching the crop grow and bear fruit. It is celebrated with fervor in many countries, certainly a time to enjoy the reaps of their bounties. It is a reverberating magic word which gives the farmer the gorgeous images of a golden field of wheat or paddy ready to be brought home. Imagine the orchards of full apple, orange and cherry trees, the fields displaying a multitude of vegetables or the vines of grapes sagging with fruit.

But however natural or noble it may be when we harvest, the act isn't all that beneficial for the crops going through the harvesting process. For the bulk of fruits and vegetables it is a painful act of separation from the mother plant. When we pluck the bunch of grapes, chop the head of cabbage, cut the hands of bananas, slash the trunk of sugar cane, pull the ears of corn, lift the roots of carrots, snap the vines of beans, how can that be beneficial? This means an abrupt termination of their life. In human law, such acts would sound grave and warrant a great many punishments. However strange it may sound when we use the same words with produce, they characterize the various changes that the produce goes through during the postharvest period. Maybe for a few crops like wheat, rice and other staple food crops, it may not look that bad since they generally stay on the plant until their fruits become fully mature and relatively dry. For most produce, harvesting marks the beginning of the deteriorative process, and the longer the crop is held (stored) before use, the lower will be its quality. Exceptions may be made in the case of fruits which attain their optimal quality following a ripening period. The various handling and storage procedures followed after the harvest period are illustrated in Figure 1.1 and determine the shelf-life of these commodities.

Because produce is highly perishable, some loss is inevitable. Depending on the produce, handling procedure, storage and environmental conditions, the extent of postharvest losses vary. Some products grown in remote areas may never even reach the marketplace, while others may have a very limited loss if they are in close proximity to their consumption. The frequently used estimate for postharvest losses of a majority of common fruits and vegetables is around 25–30%. A figure called "Food Pipeline" was published in a USDA handbook (Salunke and Desai, 1984) and depicts the physical and biological ways of occurrence of food loss from production to consumption. The commodities are shown to enter the mouth of the pipeline, and the losses through the different stages of handling and the damages due to various environmental conditions are shown as leakages from the pipe. Clearly, the quantity and quality of food filling the consumer's pot will be lower than that entering. The actual movement of food from harvest to consumer may be simpler or more complex than illustrated, but it serves to demonstrate that some loss is inevitable in every handling operations such as preprocessing, transportation, storage, processing, packaging, and marketing. A modified version of the illustration is presented in Figure 1.2.

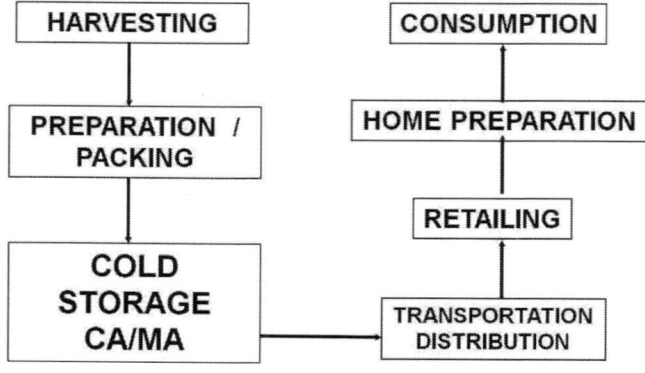

FIGURE 1.1. Typical post harvest operations.

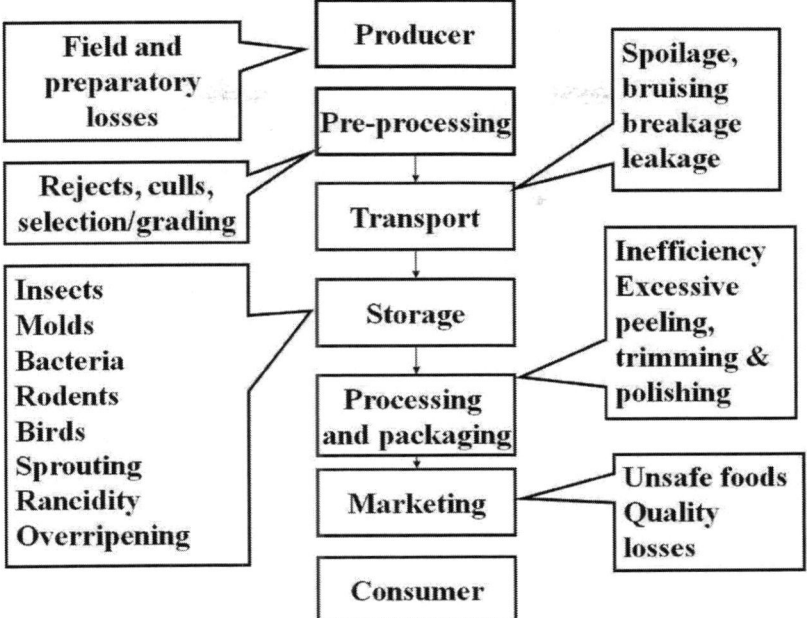

FIGURE 1.2. Losses in food chain.

Estimation of the magnitude of food losses has been the subject of considerable debate over the past several decades. Yet very little reliable information is available on the postharvest losses of perishable produce. In 1978, the U.S. National Academy of Sciences gathered some information on the postharvest losses of some commodities. These are presented in Figure 1.3 (shaded area) [data from Liberman (1981)]. Most numbers in this table are based on the "Delphi" principle, a process of summarizing the estimates (or guesses) of a number of professionals who have some standing authority in the field when hard factual data are largely unavailable. One can see that depending upon the commodity, length and conditions of storage, almost any number between 0 and 100% can be a valid estimate of the loss. One conclusion that can be drawn from this table is that it is almost impossible to quantify the postharvest losses in perishables except with reference to a particular commodity and a given location.

The above data refer mainly to nongrain commodities. The grains and other more stable food products can be expected to undergo less similar postharvest spoilage. On the other hand, since these are more durable, they are obviously stored for longer periods and one gets into types of losses other than spoilage. Infestation by insects and pests appears to be the major concern in this class of foods, and the overall loss probably is not going to be very different, especially in developing countries where storage facilities are grossly inadequate. Characteristic differences between durable and perishable food crops are detailed in Table 1.1. Notable among the differences is the low moisture content associated with the durable crops, which makes them less susceptible to microbial and enzymatic spoilage. They also are characteristically less bulky, possess harder texture and low rates of respiration.

Some data on postharvest losses in the North American scenario are provided in Table 1.2 [data from Salunke and Desai (1984); Liberman (1981)]. The estimated losses for the same commodity are at variance. For example the overall losses range from 6% to 40% from one source and they range from 2% to 30% with the other source. The total losses for apples, strawberries and peaches in one source are much higher than the magnitudes shown in the other source. The table in general indicates that softer fruits and vegetables are susceptible to greater losses, but again the magnitude depends on the commodity and the location.

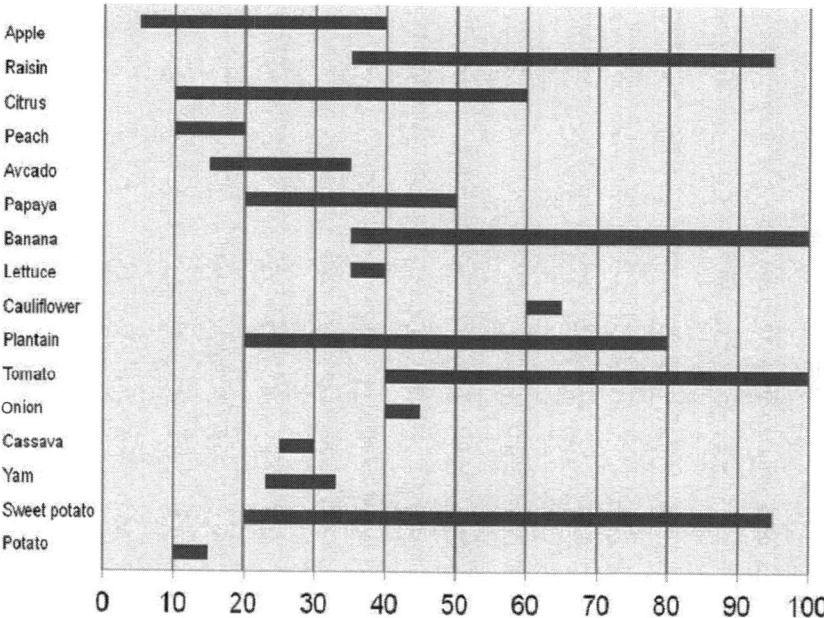

FIGURE 1.3. Range of postharvest losses in selected fruits, vegetables and staple foods.

TABLE 1.1. Characteristics of Durable and Perishable Crops.

Durables

Low moisture content, usually 10–15% or less; Small unit size, typically less than 1 g; Low-respiration rate, with very small rate of heat generation; Hard texture, not easily damaged; and Stable with natural shelf-life of several years. Losses mostly caused by external agents, e.g. agents such as molds, insects and rodents.

Perishables

High moisture content, typically 50–90%; Large unit size, typically 5 g–5 kg, occasionally even larger; High to very high respiration rate, and rate of heat production; Soft texture, easily damaged, highly perishable, with natural shelf-life of a few days to at best few months; and Losses mainly caused by external rotting by bacteria and fungi and partly by endogenous factors, respiration, senescence and sprouting.

In spite of these varying reports, one can roughly say that postharvest losses of 25% to 30% are not uncommon, especially in the absence of proper postharvest activities.

So, obviously there is a lot of wastage in the postharvest chain. What can one do about it? Shouldn't we try to reduce this wastage? Or should

TABLE 1.2. Estimated Losses of Some Fresh Fruits and Vegetables.

Produce	Percent Loss				Ref
	Wholesale	Retail	Consumer	Total	
Apples	2.9	2.9	2.4	**8.2**	a
Apples (Red Delicious)	0.2	0.2	1.5	**1.9**	b
Oranges, Navel	1.9	1.9	2.3	**6.1**	b
Peaches	12.3	5.8	10.8	**28.9**	a
Peaches	2.3	4.5	8.1	**14.9**	b
Pears, d'Anjou	–	2.5	1.6	**4.1**	b
Strawberries	13.5	5.5	22.2	**41.2**	a
Strawberries	5.9	4.9	18.0	**28.8**	b
Valencia Oranges	1.4	0.8	3.7	**5.9**	a
Cucumbers	–	5.0	2.9	**7.9**	b
Lettuce, Iceberg	4.1	4.6	7.1	**15.8**	b
Peppers, Bell	7.1	9.2	1.4	**17.7**	b
Potatoes, Katahdin	1.3	–	3.6	**4.9**	b
Sweet potatoes	–	5.7	9.4	**15.1**	b
Tomatoes, packaged	–	6.3	7.9	**14.2**	b

[a]Salunke and Desai (1984).
[b]Liberman (1981).

we concentrate on growing more food and/or intensify the agricultural production to compensate the losses? The early answer to this question was simply to grow more food to compensate for the loss since the food demand had to be met in spite of the loss. Increasing the world availability of food has been an increasing priority and it is especially so now with the world population expanding at a rate faster than food production. It is postulated that the world population will increase at a rate of 50% every 20 years which implies that the food production must increase at a simple annual rate of about 2.5% even to maintain the present day standards. Even so, that is considered grossly inadequate in many developing and underdeveloped countries. It is a matter of speculation as to how this increased production rate will be achieved, but it is generally accepted that except in a very few situations, there isn't much scope to bring new land into cultivation each year. In many countries, urbanization is actually reducing the available land for cultivation.

The second approach is to maximize agricultural production by employing modernized agricultural operations, intensified planting, high yield varieties, effective use of growth promoters, etc. These aspects were the foci of attention in the past several decades, especially in developed countries, but have limited scope in developing or underdeveloped countries where agricultural practices are mostly limited to multitudes of tiny farming sectors. Therefore, alternate measures should be adopted, including intensification of agricultural production where applicable.

One practical alternative that results in increased agricultural products availability is to minimize postharvest losses. This is a desirable alternative for many reasons. As detailed in Salunke and Desai (1984), there are several advantages with this approach:

1. *Nutritional advantage:* Since less food is lost for whatever reason, there will be much more nutritious and wholesome food.
2. *Economic advantage:* Wastage of food represents an economic loss. The economic loss increases as the food moves down the food pipeline, because to the cost of food that is lost must be added the costs of handling, transportation, storage, etc.
3. *Feedback incentive:* In some countries, the farmers could very well increase their production, but cannot store food for longer periods. So there is no economic incentive to increase production because they know that the increased output ends up as further wastage.

Reduction in postharvest losses would give them the feedback incentive to increase agricultural production.
4. *Cost-effective:* The food supply is significantly increased without bringing another acre of land into production and without using greater amounts of energy, water and capital.
5. *Environmentally friendly:* It will reduce environmental pollution and garbage disposal problems.
6. *Consumer satisfaction:* Consumers will be more fully satisfied and will receive more wholesome food.

CAUSES OF POSTHARVEST LOSSES

Although it is simple to suggest minimizing losses during various postharvest operations, achieving the goal is quite challenging. In order to do so, one must first understand the various causes of postharvest spoilage of fruits and vegetables and the factors that influence them, and secondly, use the postharvest conditions/operations that will result in extending the shelf-life of the produce.

The different causes of postharvest food losses may be broadly grouped as primary and secondary (Bourne, 1977; Salunke and Desai, 1984):

Primary Causes

1. *Biological and microbiological*: Consumption or damage by insects, pests, animals and microorganisms (fungi and bacteria).
2. *Chemical and biochemical*: Undesirable reactions between chemical compounds present in the food such as browning, rancidity, enzymatic changes, etc.
3. *Mechanical*: Spillages, damages caused by abrasion, bruising, crushing, puncturing, etc.
4. *Physical*: Improper environmental and storage conditions (temperature, relative humidity, air speed, etc.)
5. *Physiological*: Sprouting, senescence, other respiratory and transpiratory changes.
6. *Psychological*: Human aversion or refusal due to personal or religious reasons.

Many of these factors have synergistic effects, and the losses can be greater with a combination of factors. For example, chemical, microbial, biochemical, or physiological activity in a stored product is significantly influenced by the storage conditions, especially temperature. A ten-degree change in temperature can result in a two- to three-fold change in these activities. This is a key factor utilized for advantage in cold and controlled atmosphere storage applications where the produce is held at the lowest possible temperature without getting into problems of chilling injury or freezer burn. On the other hand, if proper precautions are not taken in handling and transportation, increased product temperatures may result in a very rapid quality loss.

Another example is the presence of mixed loading in a storage chamber. The emanations, especially trace gases like ethylene, from one product, may trigger deteriorative or ripening reactions in another. Again, the presence of ethylene is a factor that may have an advantage in controlled ripening chambers, but must be prevented in other situations.

Secondary Causes of Losses

Secondary causes usually are the result of inadequate or nonexistent input and may lead to conditions favorable for primary causes. This can include: improper harvesting and handling; inadequate storage facilities, inadequate transportation, inadequate refrigeration and/or inadequate marketing system.

The various causes of postharvest spoilage also can be grouped based on the nature of biological and environmental factors (Kader, 1985). The biological factors include:

1. *Respiration:* Respiration is a process by which all living cells break down organic matter into simple end-products with release of energy and CO_2. The result is loss of organic matter, loss of food value and addition of heat load which must be taken into account in refrigeration considerations. The higher the respiration rate of produce, the shorter is its shelf-life.
2. *Ethylene production:* Ethylene has a profound effect on physiological activities. Used in ripening chambers, it can trigger physiological activity even in trace amounts. Most living commodities produce ethylene as a natural product of respiration.
3. *Compositional changes:* Many changes occur during storage, some desirable and some undesirable. For example, loss of green color

is desirable in fruits but not in vegetables. Development of carotenoid pigments may have nutritional importance. There will be changes in carbohydrates, proteins, and all other food components.

4. *Growth and development:* In most produce there is continued growth and development even after harvest. Characteristic activities are sprouting of potatoes, onions and garlic, elongation of asparagus, seed germination in fruits like tomatoes, lemons, etc.
5. *Transpiration:* Transportation refers to water loss resulting in shriveling and wilting due to dehydration and is undesirable due to loss of appearance, salable weight, texture and quality.
6. *Physiological breakdown:* This includes *freezing injury* or frost damage in commodities subjected to temperatures below their freezing point which can occur in the field or during trasportation/storage. *Chilling injury* is mainly associated with tropical and subtropical commodities held for prolonged periods at temperatures between 5°C and 15°C. *Heat injury* can result in commodities exposed to direct sunlight or excessively high heat for prolonged intervals.
7. *Other factors:* These include physical/mechanical damage to the produce occurring during harvesting, handling, storage and transportation, as well as spoilage due to pathological causes (attack by microorganisms such as bacteria and fungi).

The environmental factors include temperature, relative humidity, atmospheric composition, light and other factors (fungicides, growth regulators, etc). It generally is recognized that higher temperature will result in increased respiratory activity and hence lowered shelf-life. Very high relative humidity conditions may lead to mold growth on produce surfaces while lower relative humidity can result in desiccation. Lowering of oxygen and increasing of carbon dioxide levels in storage atmospheres have been successfully used to promote microrespiration in produce and thus extend the shelf-life.

The various causes of postharvest losses will be discussed in greater detail in the forthcoming chapters, which also cover measures that can be taken to minimize these deteriorative changes. This chapter serves to provide an overview of the various causes of postharvest spoilages. Basically, one should try to minimize losses due to each and every factor in order to extend the duration of postharvest storage. The best technique would certainly involve harvesting the produce at the optimum stage of maturity, followed by quick cooling, packaging and transfer to a controlled atmosphere storage, where the temperature, relative humidity, air velocity and atmospheric composition are set at the most appropriate level for the produce in question.

The produce would be ideally left in this primary storage almost until ready for the final shipment to the retailer for quick transfer to the consumer to get at least a week's high-quality life for the produce under in-home refrigerated-storage conditions. They all sound relatively simple, but the fact that the harvesting, precooling, storage, transportation, packaging and handling requirement for each commodity can be different makes the postharvest handling system very complex.

CHAPTER 2

Classification of Fruits and Vegetables

DEFINITIONS

Fruit

FROM a botanical viewpoint, a fruit is the developed ovary of a flower; in other words, it is the product of determinate growth from an angiospermous flower or inflorescence. In a practical sense, this definition is too strict for the commodities commonly regarded as fruits by the general public. It only includes the fleshy fruits that arise from expansion of the ovary of the flower; so it does not include, for example, apples and strawberries, which arise from the growth of a receptacle, and pineapple which arises from the bract or peduncle. On the other hand, the definition includes such commodities as nuts, grains and legumes that are not commonly considered fruits. The definition also covers the common vegetables such as cucumber, tomato, peas, beans and eggplant. From a consumer point of view fruits are "plant products with aromatic flavors, which are either naturally sweet or normally sweetened before eating" (Wills *et al.*, 1989). They are essentially dessert-type foods.

Vegetable

Vegetables, on the other hand, do not represent any specific grouping from a botanical or morphological point of view, and they exhibit a wide variety of plant structures. However, based on the plant organ used, they can be much more easily grouped into several categories such as seeds, pods, bulbs, roots, flowers, buds, stems, leaves, etc. From a consumer's point of view "vegetables are soft edible plant products that are commonly salted, or at least not sweetened, cooked and often eaten with meat or fish dishes" (Wills et al., 1989). Even this definition may fall apart from the viewpoint of the fiber-conscious consumer who favors eating raw salad vegetables with or without a dip or dressing.

The derivation of some common fruits and vegetables is shown in Figure 2.1(a) and 2.1(b). It is interesting to note, for example, strawberry comes from the receptacle, pineapple from peduncle, peach from mesocarp, grape from pericarp and orange and tomato from interlocular tissues. Unlike the case of fruits, the plant part that gives rise to a vegetable is somewhat readily recognized from its appearance. Thus artichoke is a flower bud, asparagus is a stem sprout, spinach is a leaf blade and carrot is a root. Some are a little more difficult to characterize; for example, potato is a modified stem classified as a tuber, while sweet potato is a swollen root.

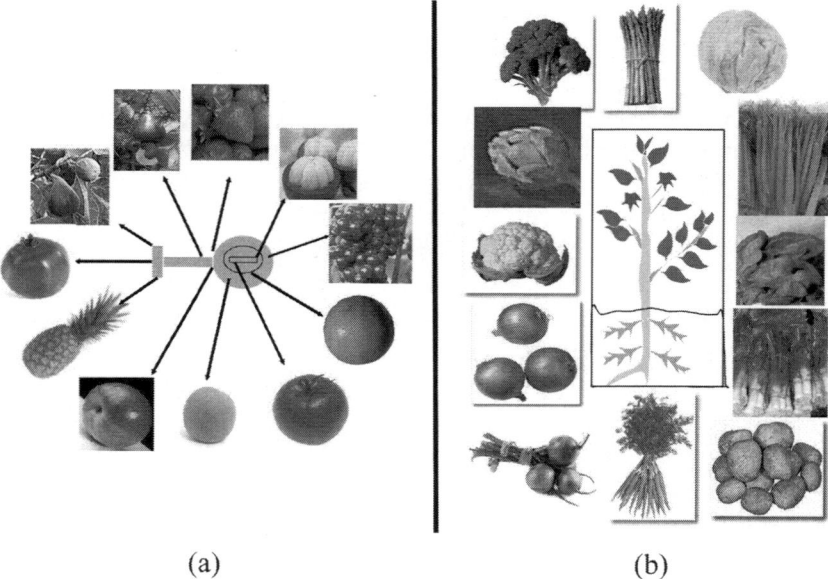

(a)   (b)

FIGURE 2.1. Derivation of some (a) fruits from different parts of the flower and (b) vegetables from different parts of the plant tissue.

CLASSIFICATION OF FRUITS AND VEGETABLES

Fruits

Although fruits naturally belong to a single large group, they may be classified on structural or morphological grounds into a several distinct types. The fleshy fruits (most of what are normally considered "fruits") are classified as Drupe, Berry, Pome, Hesperidium, Pepo, Synconium and Sorosis. The "dry fruits" which mature naturally in the dry state

✓ DRUPE: Stone Fruit - single seeded stone fruit demonstrating a simple morphological evolution

- ✓ Developed from a single carpel or from syncarpous gynoecium of a single flower
- ✓ Pericarp has a thin outer skin (epicarp)
- ✓ Fleshy middle layer (mesocarp)
- ✓ Thick hard shell (endocarp) surrounding a single seed
- ✓ Examples are cherry, peach, apricot, plum

✓ DRUPE - aggregate

- ✓ Can have also AGGREGATES of several druplets (developed from a single apocarpous flower with several individual carpels).
- ✓ Examples are raspberries and blackberries

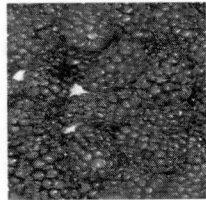

FIGURE 2.2(a). Classification of fruits (1): Drupe.

BERRY - consists of a simple morphological structure with a thin skin enclosing a juicy flesh containing many seeds

- Examples are grapes, bananas, currants, blueberries, papayas etc.

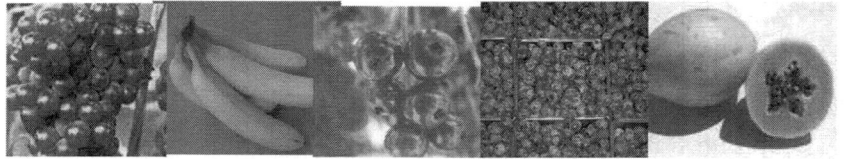

Peel structure is markedly different for each kind

POME
- Characterized by a flesh developed from the fleshy receptacle which surrounds a harder core containing seeds as in apples and pears

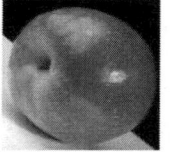

- Aggregated Pome: **Strawberries** represent an aggregate of single seeded fruits like drupes on a fleshy receptable

FIGURE 2.2(b). Classification of fruits: (2): Berry (3) Pome.

include cereals (wheat, barley, rice, etc.) and nuts are not detailed here. Legumes (pods of pea, bean, etc.), which also belong to the "fruit" group are considered with vegetables.

Some terms used in describing morphological characteristics of fruits are as follows: The individual seed bearing structures of the flower are called *carpels* and these may be separate from each other (apocarpus) or fused together (syncarpous). Collectively, they constitute *gynoecium*.

The seed-containing cavity of a carpel is called the *ovary*, the wall of which develops into the *pericarp* of the fruit. The edible fleshy part of a fruit generally develops from the ovary wall, but in some cases is derived partly or wholly from the tissues of the *receptacle*, the enlarged tip of the stem from which the floral organ arise. In some other commodities, organs such as *bracts* (leaf-like structures protecting flowers) may also enlarge and become fleshy as in the case of pineapples.

- ✓ **HESPERIDIUM covers the citrus fruits that are a modified form of berries with a well developed peelable endocarp**

- ✓ **PEPO includes fruits belonging to the cucumber family with a berry like characteristic but with a hard outer layer developed from the receptacle.**
- ✓ **Examples are melons, cantaloupe, cucumber, etc.**

FIGURE 2.2(c). Classification of fruits: (4): Hesperidium (5) Pepo.

- **SYNCONIUM represents a multiple morphological behavior. It represents a class of fruits with a hollow fleshy receptacle containing fruits derived from several individual flowers.**

- **A classical example is the fig.**

  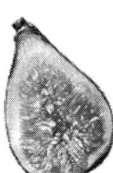

- **SOROSIS refers to the group of fruits with fleshy floral bracts and receptacle with a terminal leafy shoot.**

- **Pineapple is an example.**

FIGURE 2.2(d). Classification of fruits: (6) Synconium and (7) Sorosis.

A brief description of the different classes of fruits are given below. *Drupe* is a single seeded stone fruit demonstrating a simple morphological evolution. There can be *aggregates of druplets* developed from a single apocarpous flower with several individual carpels or several druplets compacted into a fruit. A *berry* consists of a simple morphological structure with a thin skin enclosing a juicy flesh containing many seeds. *Pome* fruits are characterized by a flesh developed from the fleshy receptacle which surrounds a harder core containing seeds as in apples and pears. The *Hesperidium* group covers the citrus fruits which are a modified form of berries with a well developed endocarp. *Pepo* includes fruits belonging to the cucumber family with a berry-like characteristic but with a hard outer layer developed from the receptacle. *Synconium* is a group representing *multiple* (collective) morphological behavior (an infructescence developed from the whole inflorescence of many flowers). *Sorosis* refers to the group of fruits with fleshy floral bracts and receptacle with a terminal leafy shoot.

The fruits belonging to the different groups are illustrated in Figure 2.2(a)–(d). A detailed description of various fruits and vegetables including their family names, scientific names and types is available in the literature. For example, Pantastico (1975) has detailed tables with above listings for selected fruits and vegetables. As an example illustration for a fruit mango, the following details are given: Family—Anarcardiaceae; Common name—Mango; Scientific Name—*Mangifera indica L.*; Type—Fleshy drupe; Description—Touch rind, extensive fleshy mesocarp with a stony outer endocarp and inner papyraceous membrane; and for the vegetable, asparagus: Family—Liliaceae; Common Name—Asparagus; Scientific Name—*Asparagus officinalis L*; Type—stem; Description—Fleshy shoot with spirally arranged scales.

Vegetables

Vegetables can be classified according to the plant organ from which they grow. Kader (1985) classified different vegetables into (1) bulky vegetative organs mostly comprising of roots, tubers and bulbs, (2) leafy succulent tissues and (3) fruit vegetables as follows:

Bulky Vegetative Organs

- *Roots:* Beet, cassava, carrot, horseradish, radish, parsnip, sweet potato
- *Tubers*: Potato, yam
- *Bulbs:* Onion, garlic

Leafy Succulent Tissues

- *Leafy:* Brussels sprouts, cabbage, celery, lettuce, parsley, spinach

- *Floral:* Artichokes, broccoli, cauliflower
- *Stem:* Asparagus, fennel

Fruit Vegetables

- *Immature fruit:* Bean, cucumber, eggplant, okra, pea, pepper, squash, sweet corn
- *Mature fruit:* Melons, pumpkin, tomato

Other Classification Methods

Classifications of fruits and vegetables have also been attempted based on various other methods. In terms of functional properties, they are classified according to their respiration rates, respiratory behavior, ethylene production rate and chilling sensitivity. The classification based on respiration rate is of considerable importance because the rate of deterioration of harvested commodities is generally proportional to the rate of respiration. The higher the rate, the lower is its keeping quality. Based on the respiration rate at 5°C, Kader (1985) grouped the fruits and vegetables into the following categories:

- *Respiration rate, < 10 mg CO_2/(kg.h):* Nuts, dates, dried fruits and vegetables; apple, citrus, grape, kiwifruit, garlic, onion, potato, sweet potato
- *Respiration rate, 10–40 mg CO_2/(kg.h):* Apricot, avocado, banana, blackberry, cherry, fig, peach, nectarine, pear, plum, raspberry, strawberry, cabbage, cauliflower, carrot, lettuce, lima bean, pepper, tomato
- *Respiration rate, > 40–60 mg CO_2/(kg.h):* Artichoke, asparagus, snap bean, broccoli, mushroom, green onion, pea, spinach, sweet corn, Brussels sprouts

If the storage-life of produce is plotted against the associated respiration rate of the produce, the resulting plot would demonstrate a steep decrease in the shelf-life as the respiration rate increases (Figure 2.3). Conversely, the figure also demonstrates that lowering of respiration rate has a tremendous potential for extending the shelf-life of produce.

Fruits are further classified as climacteric and nonclimacteric based on their respiration and ethylene production rates during development, maturity, ripening and senescence. Kader (1985) grouped the following fruits and vegetables to the climacteric and non-climacteric groups:

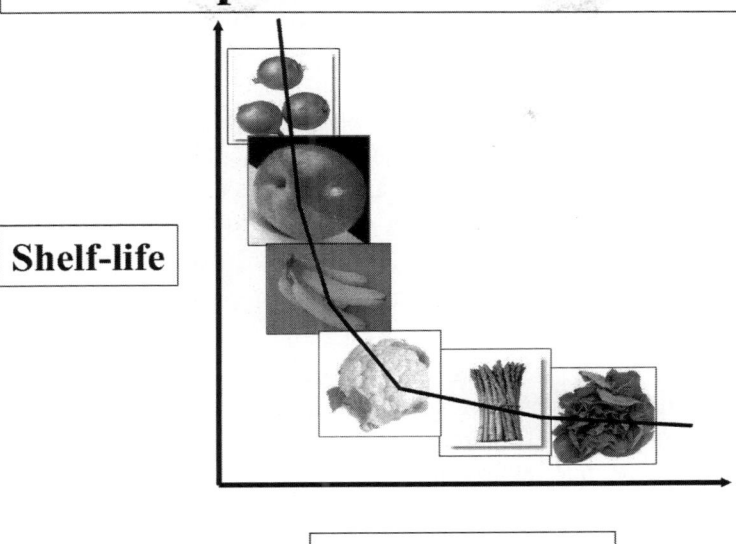

FIGURE 2.3. Storage life of produce vs their respiration rate.

Climacteric Fruits

Apple, apricot, avocado, banana, biriba, blueberry, breadfruit, cherimoya, feijoa, fig, guava, jackfruit, kiwifruit, mango, muskmelon, nectarine, papaya, passion fruit, peach, pear, persimmon, plantain, plum sapote, soursop, tomato, watermelon.

Nonclimacteric Fruits

Blackberry, cacao, cashew apple, cherry, cucumber, eggplant, grape, grapefruit, jujube, lemon, lime, loquat, lychee, olive, orange, pepper, pineapple, pomegranate, raspberry, satsuma mandarin, strawberry, summer squash, tamarillo, tangerine.

Climacteric fruits are those which exhibit a large increase in the respiration and ethylene production rates almost coincident with their ripening, while the nonclimacteric fruits show a somewhat uniform rate throughout (Figure 2.4).

Another classification based on relative ethylene production rates has a practical significance, especially when handling mixed commodities during storage or transportation. As mentioned, ethylene is physiologically very active even in trace quantities. Ethylene resulting from the use of gas-operated fork-lifts in storage chambers or produced from some commodities may trigger physiological reactions in other commodities. Such changes may be undesirable in some situations and not be objectionable in others. Fruits and vegetables can be classified into the following groups based on their ethylene production rates (Kader, 1985):

- *Ethylene production rate at 20°C below 10 μl C_2H_4/kg.h:* Banana, blueberry, cauliflower, cherry, citrus, cranberry, fig, guava, mango, honeydew melon, plantain, grape, jujube, persimmon, pineapple, pomegranate, strawberry, raspberry, tomato, watermelon, artichoke, asparagus, potato, leafy and root vegetables, cucumber, eggplant, okra, olive, pepper, pumpkin.

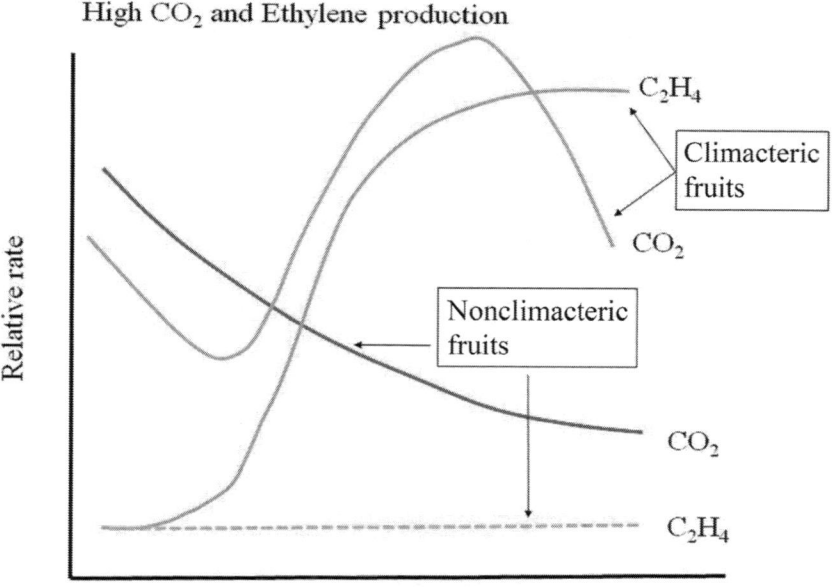

FIGURE 2.4. Climacteric behavior of fruits demonstrating steep increases in respiration and ethylene production rates at the onset of senescence.

- *Ethylene production rate at 20°C between 1–10 μl C_2H_4/kg.h:* Apple, apricot, avocado, cantaloupe, kiwifruit (ripe) nectarine, papaya, peach, pear, plum.
- *Ethylene production rate at 20°C above 100 μl C_2H_4/kg.h:* Cherimoya, mammee apple, passion fruit, sapote.

The classification based on sensitivity of the commodity to chilling injury has practical implications. Chilling injury is a common physiological disorder in tropical and subtropical commodities which are held at temperatures in the range 0 to 10°C. To prevent this, a particular commodity should never be stored at temperatures at or below which they might suffer chilling injury. According to Kader (1985), fruits belonging to Group II are the chilling-sensitive commodities.

- *Group I: Non chilling sensitive commodities:* Apple, apricot, bushberry, cherry, fig, grape, kiwi, nectarine, peach, pear, persimmon, plum, prune, strawberry, artichokes, asparagus, lima bean, beet, broccoli, Brussels sprouts, cabbage carrot, cauliflower, celery, sweet corn, garlic, lettuce, onion, pea, radish, spinach, turnip.
- *Group II: Chilling sensitive commodities:* Avocado, banana, citrus, guava, jujube, mango, muskmelon, papaya, passion fruit, pineapple, plantain, tomato, watermelon, snap bean, cucumber, eggplant, okra, olive, pepper, potato, pumpkin, squash, sweet potato.

Table 2.1 is an example of a classification based on the size of fruits and vegetables (Pantastico, 1975). The weight ranges for the different

TABLE 2.1. Sizes of Fruits and Vegetables.

Size Class	Weight Range (g)	Fruits and Vegetables
Very light	< 50	Lanzones, cashew, cherry, tamarind, garlic, strawberry, beans, peas, Brussels sprouts, olives, raspberries, blueberries
Light	50–100	Guava, passion fruit, onion, sweet pepper, tomato
Medium Light	100–250	Apricot, banana, sapote, carambola, radish, carrots, potato, eggplant, sweet potato
Medium	250–500	Mango, plantain, citrus, chayotte, cucumber, lettuce, peach, pear, pomegranate
Medium heavy	500–1,000	Avocado, cabbage, cauliflower
Heavy	1,000–5,000	Papaya, pineapple, soursop, durian, honeydew melon, squash
Very heavy	> 5,000	Jackfruit, watermelon

size classes are arbitrarily chosen and the commodities are grouped. The table as originally presented excludes several commodities such as beans, peas, Brussels sprouts, olives, raspberries, blueberries, etc., for example, in the *very light* class and several others in other classes. This type of classification may find good use in designing packages and transportation facilities.

In summary, fruits and vegetables are classified based on botanical, morphological, structural and functional differences. Each classification results from a specific objective basis and therefore is useful. From a postharvest management point of view, the functional classifications play an important role.

CHAPTER 3

Structure of Fruits and Vegetables

INTRODUCTION

IN the previous chapter, the classification of fruits and vegetables was discussed based on their morphological, structural and various functional characteristics. Fruits and vegetables demonstrate diversification with respect to their internal structure since they are derived from different parts of a growing plant. Hence, it is not surprising to expect large variations in their postharvest characteristics and storage behavior. Despite the diversity of structure among fruits and vegetables, some characterization is possible with respect to their structure. This is described in the present chapter.

As biological materials, fruits and vegetables support numerous chemical, biochemical and physiological activities during their growth and their subsequent postharvest life after detachment from the mother plant. As with other living systems, these activities are carried out in specialized cells or structural manifestations. In order to understand and appreciate the activities, we shall look at the anatomical and structural aspects first at the cellular level and later with respect to the overall system.

STRUCTURE OF A PLANT CELL

A schematic representation of the general structure of a plant cell is shown in Figure 3.1. The various components of the cell are: the cell wall, middle lamella, plasmalemma, vacuoles, cytoplasm, nucleus, ri-

Anatomy of a Plant Cell

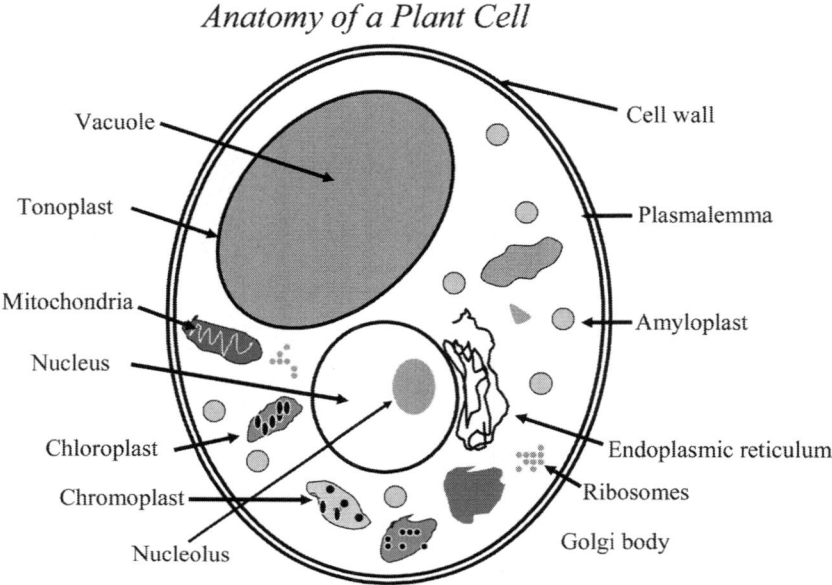

FIGURE 3.1. Schematics of a plant cell.

bosomes, mitochondria, chloroplasts, chromoplasts, amyloplasts, golgi complex, and endoplasmic reticulum. A brief outline of the functions of these components is included here. For more details, the reader is referred to more specialized textbooks on plant anatomy.

The plant cell consists of a cell wall, a semirigid structure composed of cellulose fiber and other polymeric compounds, such as pectic substances, hemicelluloses and lignins. The middle lamella also is formed from pectic substances and acts to bind adjacent cells together. The adjacent cells often have small communication channels called plasmaderma linking their cytoplasmic masses. The plasmalemma is a semipermeable membrane that actually holds all the cellular components. It is called semipermeable because it allows the free passage of water but selectively restricts or permits the movement of solutes or macromolecules such as proteins, nucleic acids, etc. Together with other semipermeable membranes like tonoplasts that surround the vacuole, the plasmalemma is responsible for maintaining the hydrostatic pressure of the cell. The resulting turgidity of the cell is responsible for the crispness of fruits and vegetables.

The main function of the cell wall is to contain the cellular components by supporting the outer cell membrane, the plasmalemma, against the hydrostatic or osmotic pressure of the cellular contents which would otherwise burst the membrane, as well as to provide structural support.

Within the plasmalemma, the cell contents comprise the cytoplasm and vacuole. The vacuole is the fluid reservoir containing various solutes such as sugars, amino acids, organic acids and salts. The vacuole is surrounded by a semipermeable membrane called the tonoplast. The cytoplasm consists of a fluid matrix of proteins and other macromolecules. Most of the life-sustaining physiological activities occur in the cytoplasm, including breakdown of carbohydrates by glycolysis for the generation of energy and various reactions leading to the synthesis of vital proteins. The cytoplasm contains several organelles which are all membrane-bound bodies with specified functions as follows:

- The nucleus, the largest organelle, is the control center of all activities. It contains the genetic information in the form of DNA (deoxyribonucleic acid) and RNA (ribonucleic acid). It is bounded by a special membrane which allows the passage of the mRNA (messenger ribonucleic acid), the transcription product of the genetic code of DNA, into the cytoplasm where it is converted to proteins on the ribosomes of the protein synthesizing system.
- The mitochondria are the powerhouse of the cell. They consist of the various respiratory enzymes of the TCA (tricarboxylic acid) cycle and house the electron transport system which synthesizes ATP (adenosine triphospate), the energy reserve of the plant system. Mitochondria utilize the products of glycolysis for energy production.
- Chloroplasts are found generally in green cells and are the photosynthetic apparatus of the cell. They contain the green pigment chlorophyll and the photosynthetic mechanism to convert solar energy to chemical energy. They also contain the necessary enzyme system for fixing of carbon dioxide for the synthesis of various carbon compounds. The chromoplast generally develops from mature chloroplast when the chlorophyll is degraded. They contain the colorful yellow and/or red carotenoid pigments. The amyloplasts are the sites of starch grain development. Collectively, the chloro, chromo and amyloplasts are called plastids.
- The golgi complex, which is a series of platelike vesicles, is considered to be important in cell wall synthesis and secretion of enzymes from the cell.

- The endoplasmic reticulum is a network of tubules and may play a vital role in protein synthesis because ribosomes are generally attached to it.

STRUCTURE OF FRUITS AND VEGETABLES

In the preceding section, we looked at the anatomical aspects of an individual plant cell with a brief description of the functions of various components. However, when we consider fruits and vegetables as a whole, we do not just look at an individual cell because produce would consist of thousands of several types of such cells. The various chemical constituents present in fruits and vegetables are distributed through a highly complex structural framework built up from the individual cellular units of a number of different kinds of tissues. Each tissue is structurally adapted to carry out certain physiological functions. Certain types of these cells are dedicated to carrying out of the various metabolic activities that keep the system alive, while some align themselves compactly at the surface to protect the inner plant tissue. Other specialized tissues are modified to facilitate the transport of various components such as water, solutes, etc., within the system. In this section, we will look at the composite structure of fruits and vegetables and discuss how the individual components add to the overall functionality.

From a functional point of view, the different tissues in a plant organ can be grouped under three main classes:

- Dermal system
- Ground system
- Vascular system

Broadly, these consist of protective, supportive and transport tissues, respectively. The following is a brief description of the different classes of tissues and their functional role in the physiological activity of fruits and vegetables during their growth, maturation and postharvest storage.

Dermal System

The dermal system primarily consists of tissues belonging to outer surfaces of the fruit or vegetable tissue. These sometimes are called protective tissues, a name that comes from their function to provide protection to the inner tissues against various deteriorative changes. The different functions of the protective tissues include:

1. Protection of produce from loss of moisture.
2. Resistance to pathogens.
3. Resistance to the penetration of chemicals.
4. Protection of produce from mechanical injuries.
5. Protection from temperature stresses.
6. Prevention of volatization of aromatic compounds.
7. Minimization of texture loss.

In addition, dermal tissues facilitate appropriate gas exchange to supply the necessary oxygen to the inner cells while at the same time getting rid of the liberated carbon dioxide.

The protective tissues are those developed at the surfaces of plant organs, forming the skins of fruits and vegetables. The cells along the outermost layer are called the epidermal cells or epidermis. The epidermal cells may vary in size and shape, but generally, they fit together compactly with almost no space between them except where breaks are provided to facilitate gas and moisture exchange. In addition, their outer tangential walls are commonly thickened and impregnated with lipidlike materials. This extracellular water impermeable surface layer on the epidermis is called the cuticle or cuticular membrane. The extent of cutinization of the epidermal cells—in other words, the nature and intensity of the coated lipid material—dictates the functional properties of the dermal system especially with reference to the resistance to moisture loss due to transpiration, resistance to the entry of pathogens and penetration of chemicals. The cuticular membrane consists of cutin or waxes. Cutin is the polymerized product of hydroxycarboxylic acids or their esters. Waxes are esters or mixtures of aliphatic wax alcohols and their corresponding fatty acids. Waxes are found in the form of granules, rodlets or platelets. Fruits and vegetables differ in their contents of waxy layers. Generally, fruits and fruit vegetables are more waxy than other vegetables, and leafy vegetables are more waxy than root vegetables.

The surfaces of underground storage organs such as roots and tubers are usually protected by a layer of corky tissue. Corky tissue also may occur in many fleshy areal organs. It is found, for example, in the skin of pome fruits where its presence removes the gloss due to the otherwise unbroken shiny cuticle. Cork cells, like epidermal cells, also are impregnated with lipidlike material called suberin.

As mentioned earlier, the epidermal layers of leaves and fruits and sometimes also of stem vegetables are interrupted by small pores called stomata, each usually surrounded by two specialized guard cells, the structure of which is characteristic to the species. The stomata, through their guard cells, can open and close, and serve as minute valves for exchange of gas with the surrounding air and also control the rate of transpiration or water loss. In the corky tissues, stomata are replaced by structures known as lenticels, in which pores are formed by the breaking apart of the cork cells. The stomatal and lenticular tissues also may provide a means of entry for plant pathogens and spoilage organisms. Therefore, between such stomatal and lenticular openings, the epidermal and corky layers form a thin continuous layer of lipidlike material.

More elaborate structural modification of the outer protective cell layers is found in seeds such as peas and beans and in the peels of some fruits such as the citrus fruits and bananas. In these cases, discrete and more or less readily separable skins are developed. The outer part of the peel, such as the flavado of citrus fruits, is also the site of numerous oil sacs into which highly flavored, essential oils are secreted.

Ground System

The ground system comprises cells where the bulk majority of metabolic activities take place and consists of Parenchyma, Collenchyma and Sclerenchyma tissues. They account for the chief edible portion of produce and also contain cells that provide mechanical support for the tissue.

Parenchyma

Parenchyma is the most common ground tissue composed largely of undifferentiated cells and is the chief edible portion of most fruits and vegetables. During development, each cell increases significantly in size, acquires a prominent, but thin, cell wall, and the major part of the internal volume becomes occupied by a single large sapfilled vacuole. The final shape of the cells is largely dependent on the interplay of forces, i.e., internal osmotic pressures and external pressure from neighboring cells during their enlargement. The intercellular contact usually is not complete with the presence of air-filled spaces in between. This void volume can vary from a low 1% as in potatoes to as much as 25% in apples. In apples, the biggest intercellular spaces are sometimes larger than the parenchymal cells. In general, the void spaces in fruits vary from 15–20%. Where there is no air, these intercellular spaces are filled with cellulose, hemicellulose or pectic materials. Most of the normal metabolic activity of the plant, including photosynthesis and glycolysis, is carried out in parenchymal tissues. In other words, the parenchymal cells house all the vital cytoplasmic or protoplasmic components.

Collenchyma and Sclerenchyma

These two types of tissues together are called the supporting tissues, and they provide the mechanical support to the organ.

Collenchyma tissue is characteristically found immediately under the epidermis, usually as a series of separate strands running longitudinally along the organ. The individual cells of the collenchyma are elongated parallel to the long axis of the organ. They are unevenly thickened with the cell walls and are especially richer in pectic materials and hemicelluloses than cellulose, which is the main constituent of other cell walls. This feature gives the walls an unusual plasticity. These tissues, although soft, are quite resilient and the strands remain intact even after cooking and mastication which cause the surrounding parenchymatous tissues to readily break down.

Sclerenchyma covers those supporting tissues in which the cells have uniformly thickened walls containing large proportions of cellulose (60–80%) and are normally lignified. The lignification process reduces the tensile strength or pliability of the cell wall, but imparts greater rigidity and hardness to provide for better impact resistance. As the tissue matures, these cells, which are primarily devoid of cytoplasm—in other words, are dead—serve only to function as a support for the plant organ. There are two types of sclerenchyma cells—the fibers which are long and pointed, and the sclereids, which, though variable in shape, usually are not very long and are not pointed. These fibers are found in closely knit longitudinal sheets and remain pretty much unchanged even after cooking. These give rise to the fibrousness or stringiness of organs such as in asparagus, beans, etc. Sclereids are especially abundant in hard structures, such as the shells of nuts and other seeds; they also occur as clusters scattered through the soft parenchymatous tissues of some fruits. Which provides a characteristic grittiness or grainy texture.

Vascular System

Vascular system tissues also are called transport tissues and mainly are responsible for the transport of water, minerals and organic products of metabolism. They consist of two principal conducting tissues: xylem, which conducts water and mineral nutrients and phloem, which plays an important role in movement and storage of organic products of metabolism and photosynthesis. Xylem forms a long open tube through which water is readily transported. Because of the presence of these long thick longitudinal cell walls, xylem tissues also provide some mechanical support to the plant organ. The phloem tissues generally stem down from the sclerenchymal tissues, and, in fact, the sclerenchymal tissues are sometimes considered as part of the first formed phloem.

The different tissues are schematically shown in Figure 3.2 (credit Encyclopedia Britanica, 1996).

In summary, the plant tissues are grouped into three classes: dermal, ground and vascular tissues. Most of the metabolic activity is carried out in the most abundant ground tissue, parenchyma, which also makes up the bulk of the fruits and vegetable tissue. The outermost layer of skin which is the epidermis—in some cases replaced by a corky tissue—is structurally modified to protect the surface of the plant organ especially with reference to gas exchange, invasion of microorganisms and attack by chemicals, etc. Mechanical support is given by the highly specialized collenchyma and sclerenchymal tissues. The vascular tis-

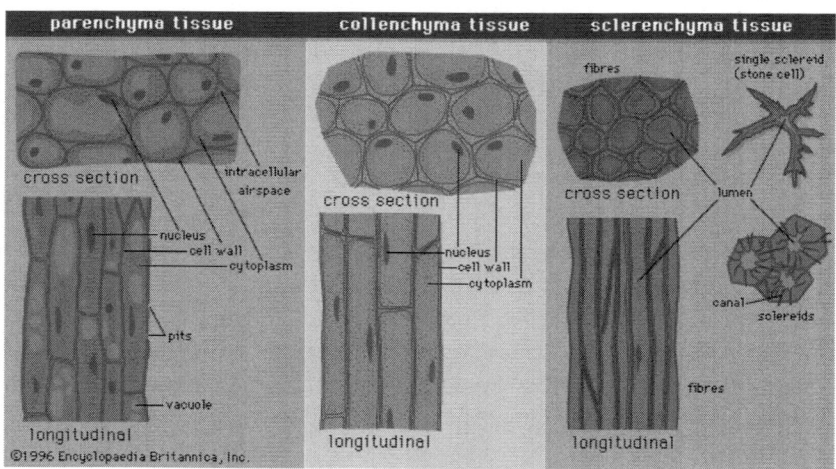

FIGURE 3.2. Structure schematics of a plant cells.

sues, xylem and phloem, which also show structural specialization assist in transport of water, minerals and organic matter from one part of the plant organ to the other.

TEXTURE OF FRUITS AND VEGETABLES

The term texture as applied to fruits and vegetables has been interpreted very broadly as to include certain features of appearance and of "hand feel" in addition to the textural characteristics experienced during the actual eating of the food (*mouth feel*). Texture is a complex property which manifests itself in many different ways. Among the more meaningful terms used to describe textural characteristics in fruit and vegetable products are: firmness, crispness, juiciness, fibrousness, grittiness and mealiness or flouriness. These attributes are dependent on the physical properties and structural organization of the main tissue constituents. The relative proportions and distribution of the various kinds of tissue, especially of thick-walled and lignified types of cells are of importance in this regard.

Firmness may be either due to the turgidity of thin-walled parenchymatous tissues or to the presence of a high proportion of thick-walled, possibly dead, mechanical tissue. Simple measurements such as resistance of the produce to a compression force may fail to distinguish between samples which in other respects would produce entirely different textural sensations. *Crispness* is a feature attributable in fresh tissues to the turgidity of the cells. *Juiciness* is generally related to the water content of the material, but since the sap initially is restricted in living tissues to the vacuoles of the individual cells, this property also depends on the extent to which the cells are burst open by teeth during mastication. The presence of discrete bundles of mechanical and/or conducting tissues which resist the shearing forces applied by the teeth gives rise to *fibrousness* or *stringiness* of texture. *Grittiness* may be caused by small particles of foreign matter or small clusters of cells with highly thickened and rigid walls (sclereids), which retain their integrity after the surrounding parenchyma breaks down. The individual cells of plant tissues for the main part are sufficiently large to be detected as separate particles by human touch. More subtle differences in texture, therefore, arise due to differences in the size and shape of the component cells and in the extent to which they become separated from each other while the food is being eaten. The separation of intact cells is a feature more commonly observed in cooked vegetables, especially those containing a high percentage of starch. Potatoes in which the cells readily separate after cooking are said to be *"mealy"* or *"floury"* in texture, while the failure of the cells to separate results in *"waxiness"* or *"soapiness."* The term *"mealy"* also has been used to describe the textural characteristic of some fresh fruits, such as apples where it is associated with dryness of the fruit surface or lack of juiciness.

Although the texture of fruits and vegetables is related to the physical and structural properties of the tissues, the mechanical model is very complex and heterogeneous. The measurement of no single physical attribute will adequately define textural quality. This does not imply that physical measurements have no value in the assessment of texture, but merely to point out that the results of any such measurement should be interpreted with caution and adequately related to other observations, for example, evaluation by a sensory panel or events related to the maturity and ripening of fruits and vegetables.

Osmotic Forces and Turgidity

The texture of fruits and vegetables depends on the turgidity of the cells. It is a result of the hydrostatic pressure within the cell moderated by the concentration of the osmotically active cellular components and the semipermeable nature of the cell wall components. The physiological processes within the living parenchymal cell enable the produce to absorb water against an osmotic pressure gradient, thus generating hydrostatic pressures that may exceed 10 atm under favorable conditions. The texture also depends on the nature of supporting tissues and the cohesiveness of the various cells. Turgor or turgidity, as this is commonly recognized, is a special word with reference to postharvest physiology of fruits and vegetables because it is almost taken synonymously to represent the freshness of the produce. No turgidity—no freshness!

Turgor, turgor pressure (TP) or turgidity is that pressure of cell contents that causes the vacuoles to enlarge and inflate against the partially elastic cell wall, thereby causing the neighboring cells to tightly press against one another to produce turgidity, rigidity and crispness to plant tissue. Vacuoles contain dissolved solids, assimilates and metabolites of various biochemical reactions, which account for the osmotic concentration of the cell.

In osmosis, a solvent moves from a region of high kinetic energy to a region of lower kinetic energy (Figure 3.3). The kinetic energy of water molecules within the cell is lowered because of the dissolved

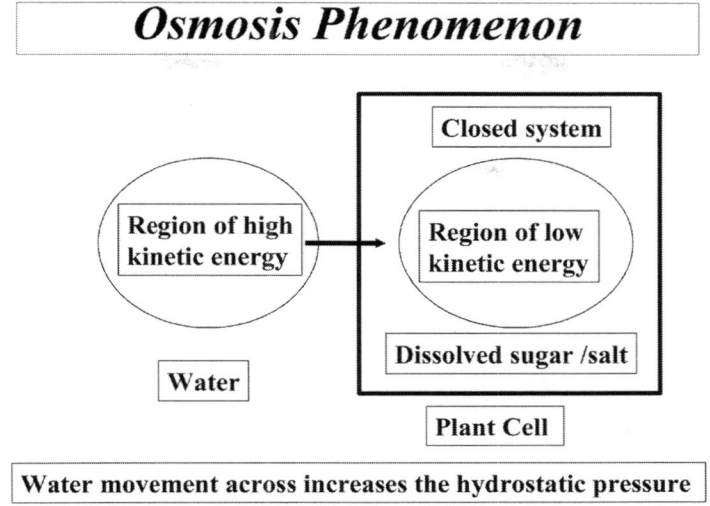

FIGURE 3.3. Schematic of the osmotic phenomenon in plants.

materials. As a result, there is an osmotic demand (osmotic potential or osmotic pressure, OP) for the water to diffuse into the cell through the semi-permeable membranes which do not allow the solutes to pass out. Diffusion of water from the plant into the cell increases the energy level of the water in the cell (in other words it dilutes the solutes and reduces the osmotic potential, OP). At the same time, hydrostatic pressure (HP) builds up and the cytoplasm inflates against the cell wall. This will lead the partially elastic cell wall to stretch, thereby creating the turgidity. Weak cells may rupture due to the increased hydrostatic pressure (HP). When the cell wall can withstand the resulting pressure, there will be a delicate balance between the hydrostatic pressure (HP) and the turgor pressure (TP), which is resistance of the cell wall to rupture; i.e., HP≃TP. The increased pressure level within the cell (i.e., reduced OP) also reduces the influx of water into the cell. In cells containing thick and strong cell walls, it is possible to have a cell wall pressure (CP) much higher than what is needed to balance the HP of the cellular contents.

As mentioned above, when the organ is attached to the mother plant, the turgor pressure (TP) will be in a *dynamic equilibrium* with the hydrostatic pressure (HP) within the cell. It is called dynamic because these pressures continuously change in response to the movement of water in the system, but at the same time the equilibrium balance is maintained. For example, when the plant cell loses water due to transpiration, this will result in a temporary increase in the osmotic pressure (OP) within the cell resulting in a pressure deficit (OP – TP), called the suction pressure (SP), which calls for the influx of water, and water is immediately drawn from the vascular tissues:

$$SP = OP - TP$$

Thus, there is no strain on the cell wall, and hence the turgidity of the cell is maintained. As more and more water is lost from the cell, OP tends to increase and then additional water will diffuse into the cells due to the osmotic demand. However, as the water accumulates in the cells, the water demand eases, OP balances with TP, resulting in lesser SP, and hence the influx of water from the plant will be minimal.

However, when the produce loses water from the cells *after it is harvested, i.e., separated from the mother plant,* OP still will increase as a result of water loss, thus creating an osmotic demand (positive SP); but the lost water can no longer be replenished by the plant. Hence, the hydrostatic pressure of the cell decreases. This will create a severe strain on the cell wall. With loss of a certain critical level of water, the HP will diminish significantly and the cell wall will begin to shrink due to the loss of volume and the tissue will collapse, giving the organ a wilted or shriveled appearance. And the turgidity of the tissue will be lost forever. It may be possible to reduce the moisture loss from the produce by storing it under a high relative humidity condition and hence delay the possibility of produce desiccation. In some rare situations, it also may be possible to reverse this situation, to some extent, if corrective actions are taken very early by produce storage under high relative humidity.

There may be some variations from the above scenario. It was recognized earlier that when the moisture from the plant cells is lost due to transpiration there will be a *positive* SP demand for the water to be replenished. In the event there is no moisture coming in, there is net loss in the hydrostatic pressure causing the produce to wilt. If the tissue is made of sturdy cells with thick cell walls, the produce may resist wilting to some extent. This means that the suction pressure created by demand of the osmotic pressure is somewhat moderated by the counter balancing cell wall pressure. Produce with thick and rigid cell walls will suffer less from the loss of moisture than produce with leafy organs. There is, however, a limit for each organ for the loss of turgidity beyond which the organ will deteriorate irreversibly. As pointed out before, temporary and smaller losses in turgidity may be reversed by

TEXTURE-STRUCTURE RELATIONSHIPS

The course ripening of fruits provides a good example of how structure greatly influences textural properties. Take for example the ripening of bananas. It is well known that the hard dark green banana turns in to a golden yellow fruit at the optimal stage of ripeness at which time it provides a soft, tender and delicious pulp. The softening of the tissues is mainly attributed to changes and the interconversion of several structurally important polysaccharides, especially to the breakdown of pectic substances and conversion of starch to sugars. The color changes can be easily observed in the photographs of bananas taken at different stages of ripening (Figure 3.4). The ripening stage of banana usually is represented by the peel color on a color chart, with a color index of 2 representing dark green stage and a color index of 4–5 representing the golden-yellow edible stage and a value of 7–8 representing the overripe stage. Figure 3.5 demonstrates the color-texture changes associated with the progressive ripening of banana. Other factors, such as cooking,

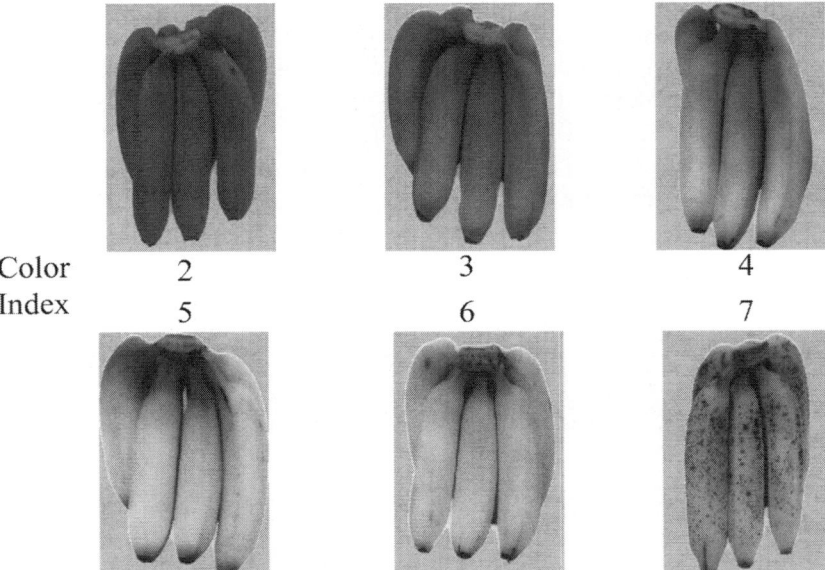

FIGURE 3.4. Color change during ripening of bananas.

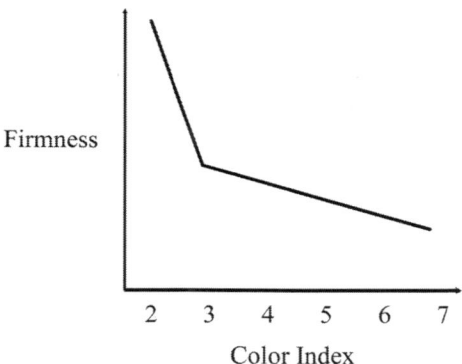

FIGURE 3.5. Texture softening related to color change during ripening of banana.

freezing, canning, drying, etc. also may bring about changes in the texture resulting in softening of the tissue. In these cases, there is usually a tissue collapse due to the treatment.

There are many ways by which texture can be measured objectively from force deformation behavior of samples using a variety of different instruments such as texture meters, universal testing machines, tensile testing equipment, etc. Often these make use of attachments for measuring puncture, compression, extension, shear, extrusion or back extrusion behavior of test samples. Some of these are illustrated in Figure 3.6.

Overall, the textural characteristics of the fruit and vegetable tissues are greatly influenced by the individual cell structure as well as their arrangements within the produce. Various instrumental techniques have been employed for measuring the food texture.

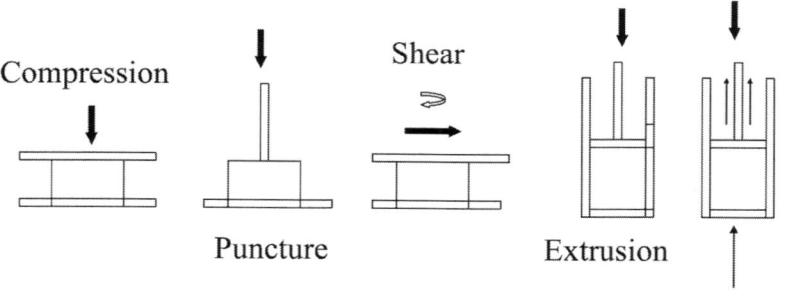

FIGURE 3.6. Various fixtures that are used for objective measurement of food texture.

CHAPTER 4

Composition and Nutritional Qualities

INTRODUCTION

PLANT foods traditionally have accounted for a major fraction of the human diet worldwide, although foods of animal origin have been recognized for their dietary contribution of proteins and key nutrients. Numerous crops fit into the dietary scene; however, some 100–200 species have been considered to be of importance in world trade, while about 15 species make up the bulk of food crops: corn, rice, wheat, sorghum, barley, sugar cane, sugar beet, potato, sweet potato, cassava, bean, soybean, peanut, coconut and banana (Fennema, 1985). The commercial groups include: cereals, pulses (legumes), root crops, stem, leaf crops, common fruits and vegetables, and a variety of plants used for extractives (oils, spices, coffee, tea, etc.).

The New Food Guide Pyramid developed by U.S. Department of Agriculture (USDA, 1995) is an excellent illustration of food choices for a balanced diet (Figure 4.1). The new graphics design is an updated version of the Food Guide Pyramid published in the July 1992 issue of Food Technology Magazine of the Institute of Food Technologists (USA) [which also could be seen on boxes of name brand cereals sold in North American supermarkets], which organizes foods into five groups and conveys the three essential elements of a healthy diet: proportion or relative amount of food to choose from each group; variety, i.e., eating a selection of different foods from each major group each day and moderation in eating of fats, oils and sugars. It recognizes that all foods are important for a balanced diet. The original food guide pyramid recommended 6–11

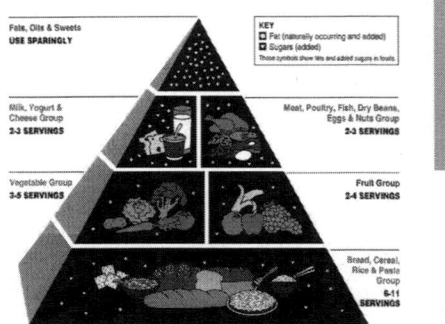

FIGURE 4.1. The Food Guide: Old and New.

servings of breads, cereals, rice, and 2–4 servings of pasta, 3–5 servings of vegetables, 2–4 servings of fruits, 2–3 servings of milk, yogurt and cheese, and 2–3 servings of meats, poultry, fish, dry beans and peas, eggs and nuts. It also recommends that fats, oils and sweets be used sparingly. The new guide calls for 6 ounces of grains, 2–5 cups of vegetables, 2 cups of fruits, 3 cups of milk, and 5.5 ounces of meat and beans. It highlights the importance of obtaining balanced amounts of protein, energy, vitamin, mineral and fiber in the diet by making proper selections from different sources.

Fruits and vegetables, like all other food materials, contain a wide range of chemical compounds and show considerable variation in their composition. A detailed discussion of the various components of individual fruits and vegetables is beyond the scope of this work. So is the detailed discussion referring to their specific chemical structure, nutritional importance, their variability with reference to commodity, variety and maturity, and the influence of various treatments and storage conditions. The reader should refer to a food chemistry book such as the one by Meyer (1960) or Fennema (1985) for such details.

The different components present in fruits and vegetables may be broadly grouped into the following major classes:

- Water
- Carbohydrates
- Proteins
- Fats
- Organic acids
- Vitamins & Minerals
- Pigments
- Flavor constituents

WATER

Water is the most abundant component of all fruits and vegetables. Water, a life-sustaining component, is as important in fruits and vegetables as in any other living system. Water also supports the activity of microorganisms, enzymes, and various other chemical and biochemical reactions taking place in the living system and hence, plays a major role in the postharvest quality and storage of the produce. Most produce contains more than 75% water. Some tissues, such as lettuce, cucumber and melons, contain about 95% water. Fruits and vegetables for the most part can therefore be considered to be expensive bundles of water in fancy packages. Some of the starchy tubers and seeds, for example, yam, cassava and potato contain lower amounts of water, but even these comprise more than 50% water. A few other plant foods like dry cereals, legumes, etc., however, contain only 10–15% moisture. Table 4.1 gives the typical composition of the different groups of fruits and vegetables (FAO Data). More detailed information may be obtained from other sources (Duckworth, 1966; Dubuc and Lahaie, 1987).

As discussed in the previous chapter, water is the most important component in fruits and vegetables from a textural point of view in terms of maintaining proper turgidity. Assuming that the water supply to the mother plant is unlimited, the moisture content in the plant organ generally attains a characteristic maximum at the state of complete turgor of the component cells. This is because, at the state of full turgor, the inflow of water from the plant to the organ is perhaps maximized (providing the highest internal hydrostatic pressure) in its attempt to satisfy the demand of the osmotic forces. There will be a delicate balance between the osmotic pressure, OP, and the turgor pressure, TP. The plant tissue under this condition generally will comprise active and healthy cells exhibiting the maximum vigor.

TABLE 4.1. Typical Percentage Composition of Edible Portion of Foods of Plant Origin.

Food	Carbohydrate	Protein	Fat	Ash	Water
Cereals	72–79	7–11	1–5	0.7–1.7	12–13
Earth—Vegetables	18–27	1–2%	0.1–0.4	1	70–78
Vegetables	2–17	1–7	0.1–0.4	0.7–1	75–95
Fruits	6–24	0.3–1.3	0.2–0.5	0.3–0.8	73–93

In very young and growing plants, the organs may contain an abundant supply of water (on a percentage basis). However, they will be very low in soluble solids. Hence, the cells exert low OP and generally will have low TP. As the organs begin to grow, they build up more soluble solids, which increase their osmotic potential and hence turgor. This will reach a characteristic maximum at their optimum stage of maturity. As fruits start to ripen, there is breakdown of starch, pectin and other cell wall materials resulting in a loss of cell wall strength, a key reason for the softening of the tissues and loss of turgor. On the other hand, as vegetables pass their optimal stage, they tend to accumulate starch at the expense of the simple sugars (soluble solids), which reduces their osmotic potential. Hence, although the cell walls continue to remain strong, the turgidity decreases due to reduction in OP. These observations also show that the moisture content of the tissue is influenced by the structural characteristics of the tissue. The accumulation of chemical constituents in various parts of the cell are detailed in Table 4.2 (Fennema, 1985).

CARBOHYDRATES

Carbohydrates are the next most abundant group of constituents. Generally speaking, approximately 75% of the solid matter of plants is carbohydrates. The carbohydrate content of fruits and vegetables vary significantly ranging from as low as 2–20% in vegetables, 5–25% in

TABLE 4.2. Structural Components of Plant Cells.

Structural Component	Chemical Constituents
Cell wall	
Primary wall	Cellulose, hemi-cellulose, pectic substances, lignin
Plasmalemma	Pectic substances, lipoproteins, phospholipids
Nucleus	Nucleoproteins, nucleic acid, enzymes
Protoplasm	
Vacuole	Inorganic salts, organic acids, oil droplets, sugars, amino acids, vitamins, pigments
Tonoplast	Pectic substances, lipoproteins, phospholipids
Chloroplast	Chlorophyll
Amyloplast	Starch granules, carbohydrates
Chromoplast	Pigments
Oil droplets	Tryglycerides
Mitochondria	Enzymes, co-enzymes, minerals
Ribosome, microsome	Nucleoproteins, nucleic acid, enzymes

TABLE 4.3. Principal Cell Wall Constituents and Their Chemical Nature.

Cellulose	β(1–4) linked glucose; Polymers of up to 12000 glucose units
Hemicellulose	β(1–4) linked xylose; Polymers of 150–200 units of xylose with uronic acid
Pectic Substances	α(1–4) linked galacturonic acid
Lignin	Insoluble, high molecular weight polymers of coumaryl, coniferyl and sinapyl alcohols
Glucomannans	β(1–4) randomly linked glucose and mannose
Arabinogalactans	β(1–3) linked galactose

fruits, 20–30% in starchy roots and tubers such as potato, cassava and yam and to more than 70% in pulses and cereals (Table 4.1). They can be present as low-molecular sugars or high-molecular-weight polymers such as starches or as dietary fiber (cellulose, pectin, lignin, etc.).

Dietary Fiber

A substantial proportion of carbohydrates in fruits and vegetables is present as dietary fiber, which is not digested by the human system because it lacks the enzyme system that can break down cellulose, hemicellulose, pectic substances, lignins and other polymers that constitute the fiber. These substances also are the major components of the cell wall. The principal cell wall constituents, their chemical nature and type of linkages and cross-linkages with carbohydrates are given in Table 4.3 (Fennema, 1985).

Decades ago, fiber was considered nutritionally unimportant in the human diet. However, today, especially in Western society, paramount importance is given to the consumption of dietary fiber. The dietary fibers have a strong affinity toward water, absorb large amounts of water and swell within the human system. Hence, they contribute nondigestible bulk to the diet and restrict the intake of other foods rich in calories and thus help control the appetite. They have been successfully used in weight control clinics. Some components of the dietary fiber are soluble in water (pectin, gums) while others are water insoluble (cellulose, hemicellulose, lignin). Foods containing dietary fiber also require chewing and they account for slower gastric emptying. It delays the absorption of glucose and fat after a meal and increases the fecal loss of bile acids, which have the capacity to bind to the fiber. Several diseases have been claimed to be due to lack of fiber in diets: appendicitis, constipation, diabetes, gallstones, obesity, hemorrhoids, tumors of rectum, etc. Consumption of fruits and vegetables in large quantities is known to alleviate such problems.

There is growing awareness of the relationship between diet and health which has led to an increasing demand for food products that support health above and beyond providing basic nutrition. Probiotics and prebiotics are components present in foods, or that can be incorporated into foods, which yield health benefits related to their interactions with the gastrointestinal tract (GI). While the benefits of prebiotics have come to light in more recent years, the importance of probiotic agents has been recognized for centuries, especially in countries where yogurt has been popular. A probiotic has been defined as "a live microbial food ingredient that, when ingested in sufficient quantities, exerts health benefits" or "live microorganisms which, when administered in adequate amounts, confer a health benefit on the host." Prebiotics are defined as "nondigestible food ingredients that beneficially affect the host by selectively stimulating the growth of one or a limited number of bacterial species in the colon, such as *Bifidobacteria* and *Lactobacilli*, which have the potential to improvethe host's health." The prebiotics simply provide support for the beneficial bacteria. Probiotic microorganisms can be found in both supplement form and as components of foods and beverages. These bacteria and yeasts have been used for thousands of years to ferment foods. Certain yogurts and other cultured dairy products contain helpful bacteria, particularly specific strains of *Bifidobacteria* and *Lactobacilli*. Prebiotics are found naturally in many foods, and also can be isolated from plants (e.g., chicory root) or synthesized. Some examples of prebiotics are inulin, fructo-oligosaccharides (FOS), polydextrose, arabinogalactan, polyols—lactulose, lactitol. Important sources of prebiotics are: whole grains, onions, bananas, garlic, honey, leeks and artichokes. In order for a food ingredient to be classified as a prebiotic, the agent should not be broken down in the stomach or absorbed in the GI tract and should be mostly fermented by the gastrointestinal microflora. Thus it has to demonstrate some selectivity in stimulating the growth and/or activity of intestinal bacteria associated with health and well-being.

Sugars and Starch

Sugars are the carbohydrate components of ripe fruits and tender vegetables while starch carbohydrates occur in vegetables and unripe fruits. The more common sugars are fructose, glucose and sucrose. Although comprising the same building blocks such as glucose units, many of the polymeric carbohydrates differ widely in their characteristics. For example, starch and cellulose, both are derived from D-glucose. While starch comprises primarily α-1,4 linkages which are easily hydrolyzed by amylase enzymes secreted in the human system, cellulose is formed by β-1,4 linkages which can only be cleaved by the enzyme cellulase, which is not present in the human system. Structures of building blocks of common macromolecules are shown in Figure 4.2. The majority of starch in fruits and vegetables is present in parenchymal tissues. They can be present in the linear (α-1,4) or branched (α-1,4 and α-1,6) polymer. Starch is used as a food ingredient in many preparations, especially as a thickening agent in soups and gravies. Industrially, starch is obtained from crops such as potato and corn. It also provides a good source for the preparation of sweet syrups, such as high fructose syrup.

Pectic Substances

Pectic substances are widely distributed in plant systems. They are major constituents of the cell wall and therefore are important with respect to produce texture. Although water soluble, pectic substances are not digested in the human system, which lacks pectic enzymes. These are long-chain polymerized methyl esters of polygalacturonic acid. Upon mild hydrolysis, these yield water soluble pectin which can form gels or viscous colloidal suspensions with sugar and acid. The gel-forming ability of pectin has been successfully used by the fruit industry to make jams and jellies. The functional properties of pectin vary depending upon chain length and the extent of methylation. Especially, the low-methoxy pectins have the capability of reacting with metal ions, particularly calcium, to form water insoluble calcium pectate or, upon cooking under appropriate conditions of pH, to form firm gels even in the absence of sugars. Again, the concept is commercially exploited in making sugarless or diabetic jellies.

The various pectic substances influence the texture of fruits and vegetables and their products in several different ways. Upon cooking, some of the water-insoluble pectic substances are hydrolyzed to water-soluble pectin. This results in a degree of cell separation in the tissues and contributes to the tenderness of cooked vegetables. Since fruits are acidic and contain sugars, the soluble pectin also tends to form colloidal suspensions, which thicken the juice or pulp of these products. In some fruit juices, the *cloudiness*, a desirable characteristic indicating fullness, comes from the colloidal suspension of soluble pectin. Starch present in vegetables also helps in thickening the products upon cooking.

Fruits and vegetables also contain a natural enzyme that can further hydrolyze pectin to the extent that it loses its gel-forming ability. This enzyme is pectin methyl esterase (PME). Several fruits, for example tomatoes, oranges, and apples, contain both pectin and the enzyme. If freshly prepared juices of such fruits are allowed to stand, the original viscosity or cloudiness gradually decreases due to the action of PME on the pectin gel. This can be prevented if the juice is heated to about 82°C which inactivates PME, an essential treatment given to fruits either during or after crushing. In contrast, when clear juices are desired, as in the case of clarified apple juice, the enzyme activity is allowed to take place

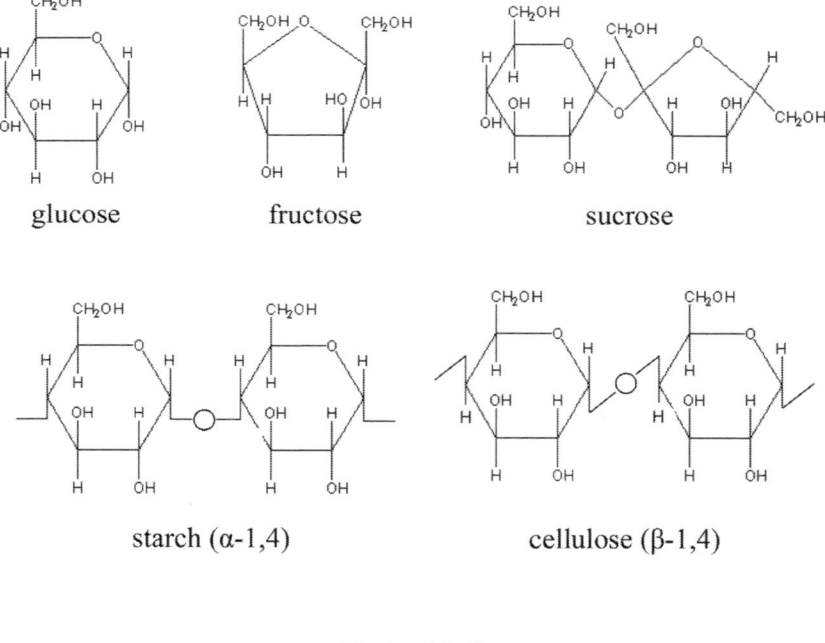

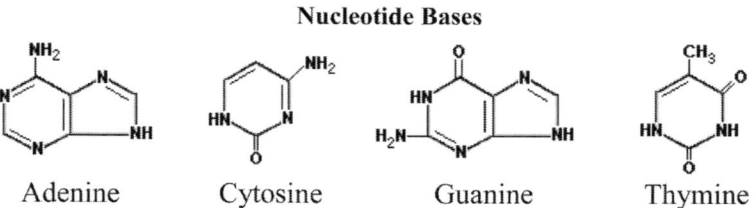

FIGURE 4.2. Basic building blocks and linkages of some macromolecules.

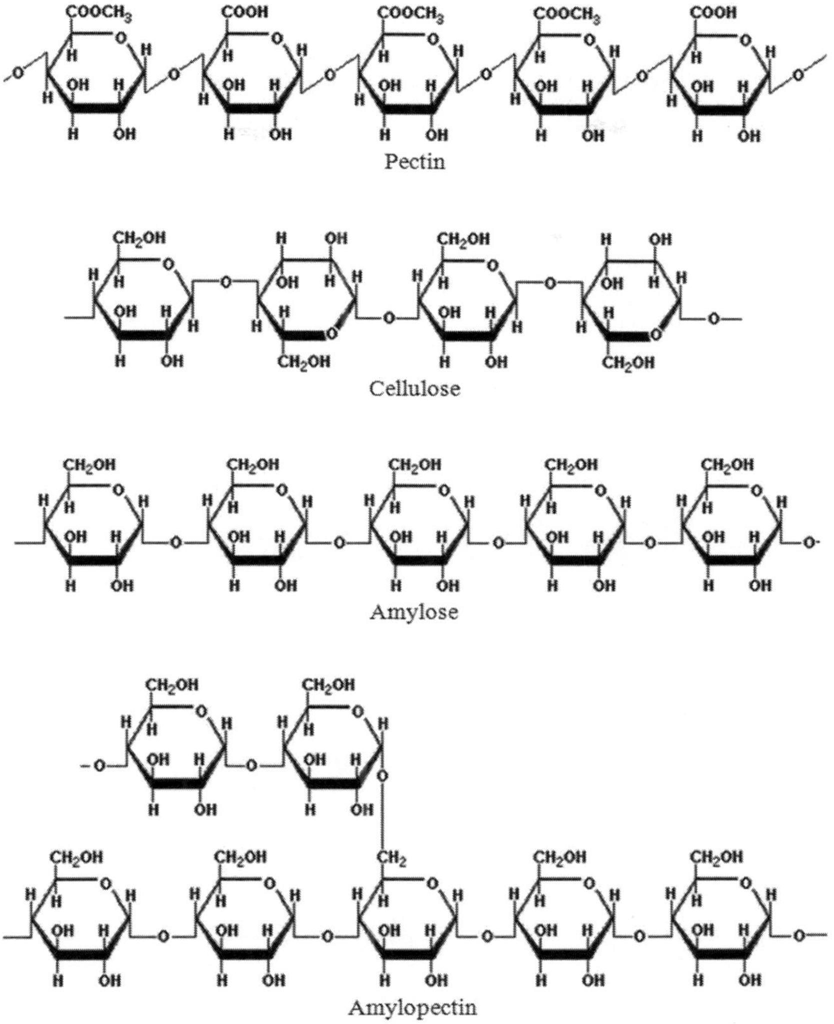

FIGURE 4.3. Structural representation of some macromolecules.

initially and the residual enzyme is inactivated by heat, and the juice is filtered subsequently. The treatment obviously improves the efficiency of the filtration process.

Prior to several preservation techniques such as freezing and canning, it is desirable to firm the texture of fruits and vegetable products which are normally softened by the processing treatment. For this purpose, the reaction between soluble pectin and calcium ions is used to advantage. A brief dip of the product in calcium chloride solution or addition of some calcium salts to food preparation has been shown to be effective in firming certain produce by forming calcium bridges.

PROTEINS

Proteins are a functionally important class of compounds, but fruits and vegetables are not major contributors of proteins. The protein content of fruits is generally less than 1% and that of common vegetables is about 2% (with the exception of peas which have about 7% protein). Cereals and pulses contain 7–12% protein. Nevertheless, they should be considered as a part of the structural constituents since they are the major components of the cytoplasm of the living cell and therefore are functionally important. Most proteins present in fruits and vegetables are "storage proteins," although they are not necessarily synthesized in plants for the purpose of storage. These macromolecules function to lower the osmotic potential of the cells because their synthesis helps to lower the content of amino acids, which are the building blocks of proteins. Although present in smaller quantities in fruits and vegetables, they nevertheless improve the nutritive value of the crop. Common plant proteins are albumins, globulins, gliadins and glutenins. Gluten is a complex formed from gliadins and glutenins following hydration and mixing, and is responsible for the rubbery texture of wheat dough.

FATS AND OILS

The lipids of fruits and vegetables (with the exception of avocado and olive) are, like proteins, largely confined to cytoplasmic layers and their content is generally less than 2% (dry weight basis). Some exceptions are palm kernel (70–80% oil), avocado (35–70% oil), olive (20–40% oil), and laurel (25–55% oil). Other oil-bearing plant crops, generally with more than 50% oil on a dry weight basis, include the various oil seeds like soybean, sunflower and sesame seeds, and several nuts (coconut, peanut, and almond).

Lipid materials are especially prominent in the protective tissues at the surface of the plant organs such as cuticle, epidermis and corky tissue. Fats are triglycerides, i.e., fatty acid esters of glycerol. The nutritional value of a fat/oil depends on its unsaturation (presence of one or more double bonds). A common advice today is to avoid or reduce as far as possible saturated fats and oils. Coconut (45% lauric and 20% myristic) and palm kernel (55% lauric and 12% myristic) oils are rich in saturated fatty acids while corn (44% oleic and 48% linoleic, 8% saturated), peanut (51% oleic and 26% linoleic, 15% saturated), olive (75% oleic and 10% linoleic, 15% saturated), soybean (20% oleic and 64 lenoeic, 13% saturated) and sunflower (31% oleic and 57% linoleic, 10% saturated) oils are rich in polyunsaturated fatty acids (mostly oleic and linoleic, small amounts of linolenic acids). Lipid content of selected foods and the fatty acid composition of selected oils are shown in Table 4.4.

The four C18 (eighteen carbon atoms) fatty acids differ with respect to their degree of unsaturation: stearic acid (fully saturated), oleic acid (one double bond), linoleic acid (two double bonds) and linolenic acid (3 double bonds). They also are identified based on the position of double bonds and their conjugation. The structural configuration of linoleic acid is written as:

$$CH_3CH_2CH_2CH_2CH_2CH=CHCH_2CH=CHCH_2CH_2CH_2CH_2CH_2CH_2COOH$$

which is abbreviated as (C18:2, 9–12):

$$CH_3(CH_2)_5CH=CH-CH=CH(CH_2)_7COOH$$

Double bonds are said to be "conjugated" when they are separated from each other by one single bond, e.g., ($-CH=CH-CH=CH-$). The

TABLE 4.4. Lipid Content of Some Foods.

Source	Lipid Content (% dry weight basis)	Source	Lipid Content (% dry weight basis)
Oil-palm	70–80	Snap beans	1.1
Avocado	35–70	Corn	2.6
Olive	30–70	Almonds	58
Laurel	25–55	Cashew	46
Grape	0.2	Coconut	65
Banana	0.1	Peanut	49
Apple	0.06	Sesame	50
Artichoke	1.3	Sunflower	50
Potato	0.1	Side salad	0.1
Potato chips	14	Side salad with dressing	1.2
Onion	0.1	Eggplant	0.2
Fried onion	3.5	Fried eggplant	7.4

term "conjugated linoleic acid" refers to several C18:2 linoleic acid variants such as 9, 11, 10, 12 etc. The 9–11 conjugated linolenic acid is:

$$CH_3(CH_2)_5CH=CH-CH=CH(CH_2)_7COOH$$

Omega 3 and Omega 6 Fatty Acids

Omega-3 fatty acids (also called ω-3 fatty acids or *n*-3 fatty acids) are fats commonly found in marine and plant oils. They are *polyunsaturated fatty* acids (PUFA) with a *double bond* (C=C) starting after the third carbon atom from the end of the carbon chain. The fatty acids have two ends—the acid (COOH) end and the methyl (CH$_3$) end. The location of the first double bond is counted from the *methyl end*, which also is known as the omega (ω) end or the n end. Some of the potential health benefits of omega-3 fatty acids supplementation are controversial. They are considered *essential fatty acids,* meaning they cannot be synthesized by the human body but are vital for normal *metabolism.* Though mammals cannot synthesize omega-3 fatty acids, they have a limited ability to form the long-chain omega-3 fatty acids, including *eicosapentaenoic acid* (EPA, 20 carbons and 5 double bonds), *docosahexaenoic acid* (DHA, 22 carbons and 6 double bonds) and *α-linolenic acid* (ALA, 18 carbons and 3 double bonds) (Figure 4.4). Common sources of omega-3 fatty acids include *fish oils* (salmon, mackerel, trout), *algal oil, squid oil,* and some *plant oils,* such as *echium oil* and *flaxseed, walnut* and *pumpkin seed oils* (Wikipedia).

***Cis* vs *Trans* Fatty Acids**

Double bonds bind carbon atoms tightly and prevent rotation of the carbon atoms along the bond axis. This gives rise to *configurational*

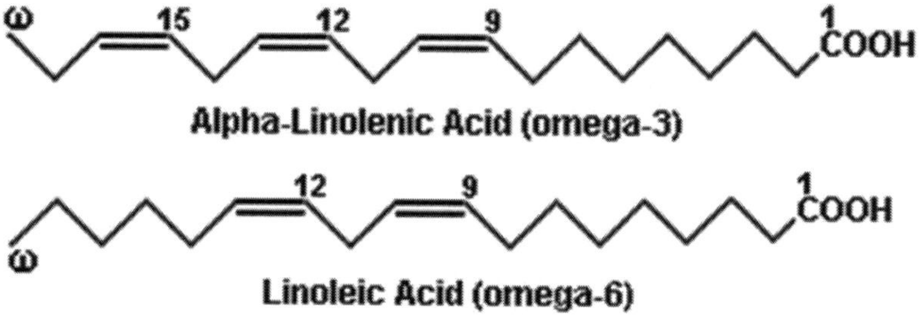

FIGURE 4.4. Omega-3 and Omega-6 fatty acids.

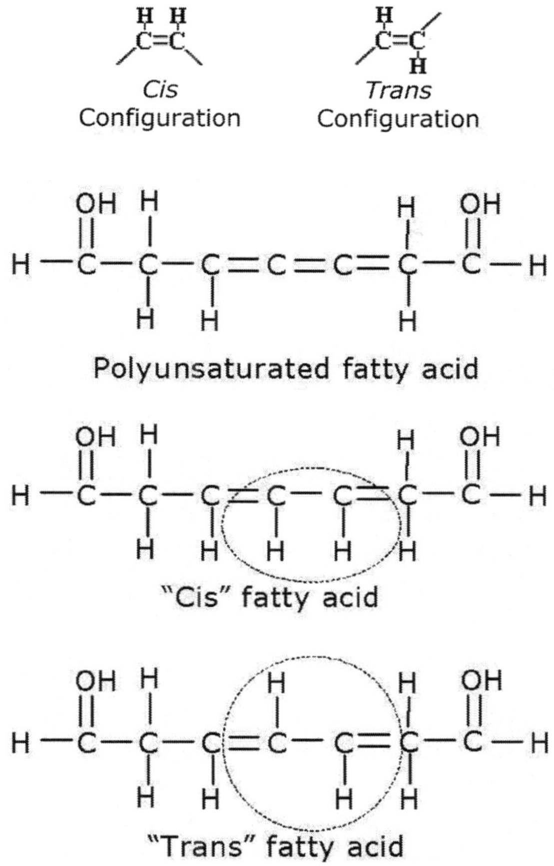

FIGURE 4.5. Cis vs Trans fatty acids.

isomers which are arrangements of atoms that can only be changed by breaking the bonds. Naturally occurring fatty acids generally have the *cis* configuration. The natural form of oleic is found in virgin olive oil and has a "V" shape due to the *cis* configuration. The *Trans* configuration (elaidic acid) looks more like a straight line (Figure 4.5). *cis* means "on the same side" and *trans* means "across" or "on the other side" (Latin). While the natural *cis* form is healthy, the trans form, which is generally formed during processing, is unhealthy and should be avoided when possible. Foods that usually contain high levels of trans fatty acids are: pastries and cakes, French fries (unless fried in lard/dripping), doughnuts, cookies/biscuits, chocolate, margarine, shortening, fried chicken, crackers and potato chips.

Dietary *trans* fats raise the level of low-density lipoproteins (LDL or "bad cholesterol") increasing the risk of coronary heart disease. *Trans* fats also reduce high-density lipoproteins (HDL or "good cholesterol"), and raise levels of triglycerides in the blood. Both of these conditions are associated with insulin resistance, which is linked to diabetes, hypertension and cardiovascular disease.

ORGANIC ACIDS

Organic acids are acids that result from the synthetic activities of plants. Most fruits and vegetables contain organic acids at levels in excess of that required for the operation of the TCA cycle and other metabolic pathways. The excess amount generally is stored in vacuoles. Organic acids naturally in fruits and vegetables are many: acetic, citric, malic, oxalic, etc. In the ripening of fruits, some of the acids also are progressively utilized in the formation of ethers and carbohydrates. Others are combined to form salts of potassium, sodium, calcium, magnesium, etc. The dominant organic acids present in produce are citric and maleic acids. Other acids dormant in some commodities are tartaric acid in grapes and avocado, oxalic acid in spinach and isocitric acid in blackberry. Lemons contain in excess of 3% organic acid and lemon juice is a common food acidulant. In many foods the acid exists in minute percentages; in none of them is the percentage of acid very great. They have a very pleasing flavor.

Citric acid is a weak organic acid with the formula $C_6H_8O_7$. It is naturally present in many citrus fruits like lemons, oranges, limes, grapefruits, etc. Lemons and limes are used as natural acidulants in most food preparations. They also are used as a natural preservative. In biochemistry, the conjugate base of citric acid, citrate, is important as an intermediate in the citric acid cycle, which occurs in the metabolism of all aerobic organisms. Citric acid is a commodity chemical, and more than a million tonnes are produced every year by fermentation. It is used mainly as an acidifier, as a flavoring, and as a chelating agent (Wikipedia).

Oxalic acid is an organic acid with the formula $H_2C_2O_4$. It is a colorless crystalline solid that dissolves in water to give colorless solutions. It is classified as a *dicarboxylic acid*. In terms of acid strength, it is much stronger than acetic acid. Oxalic acid is a *reducing agent* and its *conjugate base*, known as *oxalate* ($C_2O_4^{2-}$), is a *chelating agent* for metal cations (Wikipedia). Oxalic acid occurs widely in plant products. Oxalic acid is found in large amounts in cranberries and spinach; it is not easily digested in the human system and excess consumption has been implicated in kidney disorders. Calcium oxalate occurs abundantly in spinach and broccoli, but not in levels dangerous to humans.

Acetic acid is found in many plants. It combines readily with sodium, potassium, ammonium and other alkalis, forming salts or acetates; these acetates exist naturally in the juices of many vegetables. The acid and its salts are converted into alkaline carbonates in the body. Malic acid is found in apples, apricots, cherries, cherimoyas, currants, loquats, mangos, papayas, pears, peaches, pineapples, plums, prunes, quinces, tomatoes, blackberries, cranberries, raspberries, strawberries, either in the free state, or in combination with alkaline bases, as malates, such as malate of calcium, malate of potash, malate of magnesium; and also is found in parsley, carrots and potatoes.

Tartaric acid is one of the most common organic acids. Grapes, mangos and tamarinds and other fruits contain this acid. As grapes ripen their tartaric acid disappears and sugar and other carbohydrates increase. The acids apparently are converted into sugars and starches.

VITAMINS AND MINERALS

Fruits and vegetables are the richest source of vitamin C and virtually all of man's entire vitamin C requirement is obtained from fruits and vegetables. They also may be important sources of vitamin A and folic acid. The vitamin A is mainly present as pro-vitamin A, which is β-carotene, a carotenoid pigment, and is converted to vitamin A within the human body. Other vitamins generally are present in small quantities. Fruits and vegetables are major contributors of vitamins to the U.S. diet. For example, fruits account for 35% of the vitamin C requirement while also contributing vitamin A. Vegetables contribute more than 60% of vitamin C, 40% of vitamin A and 10–15% of thiamin, riboflavin and niacin. Cereal flours and dry beans meet 40% of thiamin, 30% of niacin and 15% of riboflavin requirements.

One should note also that the nutritional importance of various fruits and vegetables not only depends on the concentration of the nutrients in the produce, but also on their bio-availability and on the amount of such produce consumed in a typical diet. Based on the relative nutrient concentration of major vitamins and minerals, the vegetables were graded from top to bottom in the following order (1–33): Broccoli, spinach, Brussels sprout, lima bean, pea, asparagus, artichoke, cauliflower, sweet potato, carrot, sweet corn, potato, cabbage, tomato, banana, lettuce, onion, and orange, with broccoli taking the top honors and orange the last. Tomatoes and oranges were relatively low in concentration of nutrients, but they moved to rank No. 1 and 2 in terms of contribution of the nutrients to diet because of their large per capita consumption. The order with respect to the vegetables as major contributors is: Tomato, orange, potato, lettuce, sweet corn, banana, carrot, cabbage, onion, sweet potato, pea, spinach, broccoli, lima bean, asparagus, cauliflower, Brussels sprout, and artichoke. Potato moved from rank 14 to 3.

Fruits and vegetables also contribute at nutritionally significant levels of the minerals iron and calcium. Some examples of mineral rich vegetables are: Spinach (Calcium, 3–300 mg/100 g), Sweet corn (Magnesium, 2–90 mg/100 g), Seeds and sprouts (Phosphorous, 7–230 mg/100 g), Celery (Sodium, 0–124mg/100g; Chloride, 1–180 mg/100 g), Parsley, green foods (Iron, 0.1–4 mg/100 g) and High protein tissues (Sulphur, 2–170 mg/100 g). However in some instances, although these are present in large quantities, the form in which they occur may not be of nutritional significance; for example, the calcium in spinach is present as calcium oxalate which is not absorbed by man.

In recent years, there has been considerable interest on the role of calcium on the postharvest storage life of fruits and vegetables. Higher levels of calcium achieved by agronomic practices or by postharvest treatment have been shown to improve the textural and storage quality of the product. It has become increasingly evident that the mineral content of a given species can have a significant influence on the physiological disorders that arise both pre- and postharvest. The nutrients of major importance in pre-harvest application are nitrogen (N), phosphorous (P) and potassium (K), which are the major components of most fertilizers. Other minerals of significance include: calcium (Ca), magnesium (Mg), zinc (Zn), boron (B) and copper (Cu). While the extent of fertilization generally determines the crop growth, it is prudent to use supplements only when soil is deficient. It is important to have a proper balance of nutrients because both shortage and abundance can be bad for plant growth and postharvest quality of the produce. Some examples

TABLE 4.5. Mineral Imbalances and Associated Post-harvest Defects.

Mineral Application	Post-harvest Quality Defects
low K and high N	Stem end breakdown (citrus); tip burn (cabbage); cracking (sweet potato)
high P	Pencil stripe in celery
low Cu	Citrus "dieback" (dark brown rind)
high N and P	Aggravated citrus "dieback"
low B	Brown fleshy core (cabbage); blisters and cracks (cauliflower), internal black spots—beets
low Ca and high K	Cavity spots in carrots

of postharvest quality defects due to imbalanced mineral nutrition are given in Table 4.5.

PIGMENTS

The natural coloring in fruits and vegetables depens on a large number of individual chemical compounds, which fall naturally into three main groups:

- chlorophylls
- carotenoids
- flavonoids (anthocyanins)

Chlorophylls

Chlorophylls are the normal green pigments of plants and play a vital role in their photosynthetic activity. These are located in chloroplasts as "chlorophyll a" and "chlorophyll b." These two are structurally very similar; "a" has a methyl group on carbon 3 while "b" has an aldehyde group on carbon 3 of a tetrapyrole skeleton with magnesium at the center (Figure 4.6). The chlorophylls are very unstable compounds and undergo considerable change. Some pathways for the degradation of chlorophylls are shown in Figure 4.7. Notable among them is the ease with which the Mg ion is removed by heating in acid solution turning the green color to olive pigments called pheophytin and subsequently to the brown pheophorbide (upon removal of a phytol group). Enzyme chlorophyllases act on chlorophylls, converting them to bright green chlorophyllins, a process which has been commercially exploited to give better color to some green vegetables like peas, beans, etc.

Chlorophyll a: R = CH$_3$
Chlorophyll b: R- CHO

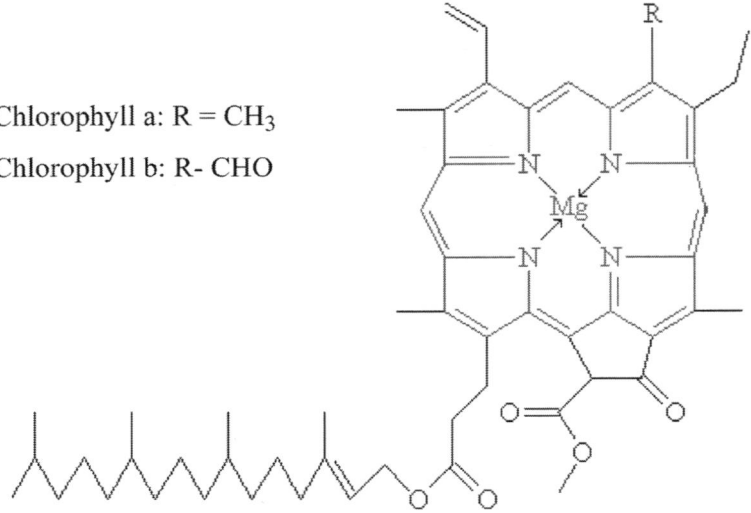

FIGURE 4.6. Structural representation of chlorophylls.

Carotenoids

Carotenoids are a group of yellow, orange or orange-red fat soluble pigments widely distributed in nature. In green leaves they occur in chromoplasts. The green color of chlorophylls generally masks the yellow to red carotenoid pigments. They are the rich pigments of many fruits and vegetables: carrots, sweet potatoes, peaches, banana skins, tomatoes, red peppers, squash etc. When consumed by animals, they tend to concentrate in lipids and hence are found in blood, milk, egg yolk etc. They include β carotene (carrot) [α and γ are two other isomers], lycopene (tomato) and xanthophylls (radish). β carotene and lycopene differ only in the cyclization of the end carbons (Figure 4.8). β carotene is a precursor for the synthesis of vitamin A in the body.

Flavonoids

Flavonoids consist of a group of naturally occurring red, blue or purple anthocyanin pigments. The shades of these pigments are pH dependent with alkaline pH favoring blue while acidic conditions bring out the red. They consist of anthocyanins (red, blue or purple pigments), anthoxanthins (yellow pigment) and catechins and leucoanthocyanins (tannins) (naturally colorless but form brown pigments).

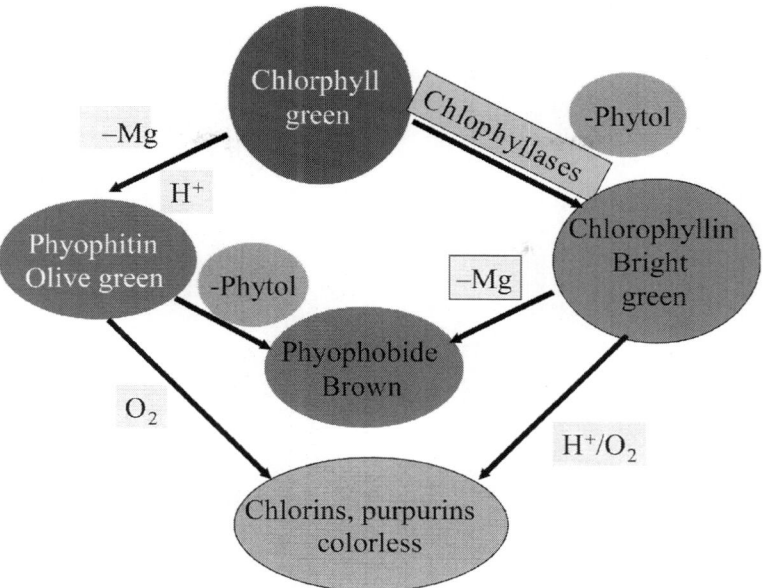

FIGURE 4.7. Some pathways for the degradation of chlorophyll.

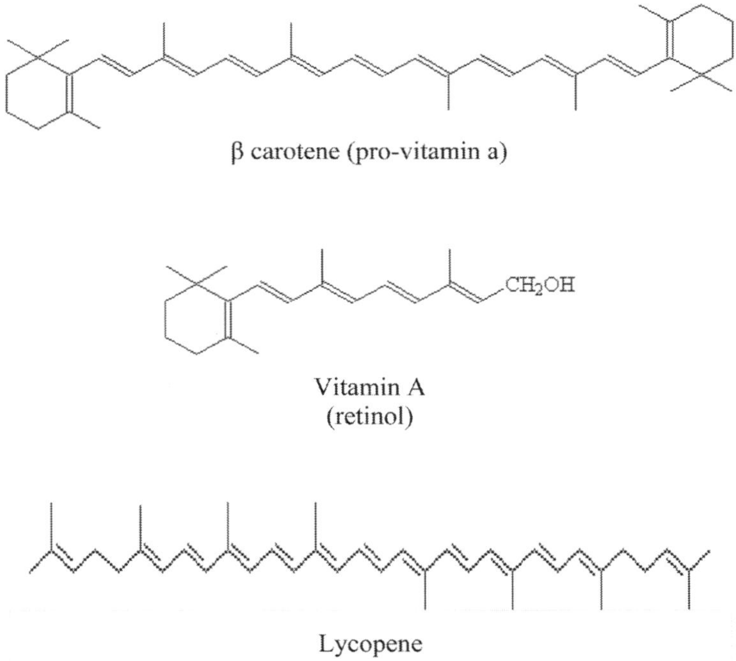

β carotene (pro-vitamin a)

Vitamin A
(retinol)

Lycopene
(no provitamin activity)

FIGURE 4.8. Structure of β carotene, vitamin A and lycopene.

Anthocyanins

Shades of anthocyanins are pH dependent. They are red under acidic conditions (low pH), slowly change with increase in pH to violet and finally to blue under alkaline pH. Common sources of anthocyanin pigments are cherries, strawberries, blueberries, cranberries and eggplants.

Anthoxanthins

Anthoxanthins are yellow- to orange-colored pigments occurring in the sap of fruits, stems, leaves, roots (orange and lemon). They have nutritional/pharmaceutical importance, reported to be active components in some drugs (vitamin P) and are used as an anti-hemorrhage factor.

Tannins

Tannins are recognized historically as tanning agents which have the capability of reacting with components of skin and "tan" them. Tannins react with metal ions to form dark colors. Commercially, they are used in making inks. The substances that react with proteins of skin to produce tan are polymers of catechin. The dark color of coffee and tea are from tannins which gives body and fullness to these beverages. They also cause astringency in foods (tea, coffee, hops).

FLAVOR COMPOUNDS

All fruits and vegetables produce a range of low-molecular-weight compounds that possess some volatility at ambient temperatures while some form these after cooking. These compounds are important in producing the characteristic aroma of fruits or cooked vegetables.

TABLE 4.6. Some Typical Flavor Components.

Apple	ripe: ethyl 2-methylbutyrate; green: Hexanal, 2-hexenal
Banana	green 2-hexenal; ripe: eugenol; overripe: isopentanal
Lemon	citral
Orange	valencene
Cucumber	2,6-nonadienal
Potato	2-methoxy-3-ethyl pyrazene, 2,5-dimethyl pyrazine
Radish	4-methylthio-trans-3-butenyl isothiocyanate

ANTI-NUTRITIONAL FACTORS IN FOODS

The discussion of composition and nutritional qualities of fruits and vegetables would be incomplete without mentioning the anti-nutritional factors, which may be naturally present in some fruits and vegetables or find their way in through food additives, microbial activity, or contaminants. It is well recognized that in nature there exist plants with the capacity of synthesizing a multitude of chemicals that cause toxic reactions when consumed by humans and animals. Historically, man also has learned to avoid plants that cause acute, easily recognizable poisoning or to develop processing methods that reduce or eliminate the toxicity. However, if consumed without discretion, the food we eat can induce neurotoxicity, such as allergies, neurolathyrism, Parkinsonian dementia, loss of memory, convulsions and even death.

In some parts of the world (southern parts of India, for example) a wild sweet pea called *Lathyrus sativus* consumed by poor people, especially during floods or drought periods, has been linked with paralysis and difficulty in walking. It was discovered that it is due to the presence of an active constituent chemically named β-n-methylaminoalanine, a structural analog of glutamate. Monosodium glutamate (MSG) is used in various dishes prepared in Chinese restaurants has been associated with Chinese syndrome (allergies, body rashes and burning sensations). Scientific studies have shown recently that increased release of glutamate in the brain may induce excitotoxicity to cause neuronal death and hence impairments in learning, intelligence, memory and behavior. In certain regions of the world such as Fiji, Guinea, Guam and Mexico, the bark of a plant called *Cycus circinalis* is used to prepare bread. This contains another important neurotoxic ingredient called β-n-alylaminoalanine, which induces *Parkinsonian dementia*. Mushrooms generally are consumed all over the world. However, certain wild varieties of mushrooms are highly toxic and even fatal if consumed without quality control. The active ingredient of mushroom is called acromelic acid, which can induce symptoms of amyotrophic lateral sclerosis (ALS) (producing spinal cord damage). Recently, about 250 people got sick and some of the elderly died by the accidental ingestion of domoic acid contaminated mussels on Prince Edward Island (Canada). Domoic acid like kainic acid is a rigid structural analog of glutamic acid and selectively damages the hippocampal dentate gyrus and CA3 layer neurons involved in memory storage mechanisms. Domoic acid con-

TABLE 4.7. Toxic Constituents of Plant Foodstuffs.

Name of the Toxin	Occurrence in Food
Protease (Enzyme) Inhibitors	Beans (soy, mung, kidney, lima, navy), chickpeas, peas, potato cereals
Hemaglutinins	Beans (soy, kidney, black, yellow), peas, lentils
Saponins	Soybean, peanuts, sugar beets, peanuts, spinach, asparagus
Glucosinolates (goitrogenic)	Cabbage, turnip, rutabaga, radish, rapeseed, mustard
Cynides and cynogens	Peas, lima beans, pulses, flax, cassava, tapioca, almonds
Gossypol	Cottonseed
Allergens	Practically all foods
Phytoalexins	Sweet potato, celery, parsnip, broad beans
Alkaloids	Mushrooms, herbal tea
Solanin	Sprouting potatoes
Aflatoxin	Peanuts
Furocumarin	Parsnip
Oxalic acid	Rhubarb, spinach

taminated mussels were eaten as salad and soup by individuals, which induced permanent loss of their memory.

Traditional doctors of Japan used to provide the decoction of the seaweed, called *Digenea simplex*, which contained kainic acid, to children infested with helminths in their intestine, which induced neurobehavioral symptoms and memory impairment in children when they reached adolescence. Similarly many of us already are aware that soybean contains thermolabile toxic peptides (protease inhibitors, hemagglutinins etc.) which have to be removed by heating at 60°C before cooking. Excessive consumption of vegetables of the cabbage family (contains glucosinolates) can lead to hypothyroidism and thyroid enlargement; some peas, beans and pulses contains cyanogen, which can lead to HCN poisoning; raw eggs contain avidin which if not inactivated by heat can interfere with vitamin absorption. Likewise there are many other examples in the health literature. Table 4.7 lists most of the major toxicants found in plant foods along with descriptions of their essential features. It is accurate to say that on scientific grounds both vegetarian as well as non-vegetarian food can be harmful (even fatal) if ingested without proper awareness and quality control.

CHAPTER 5

Harvesting of Fruits and Vegetables

INTRODUCTION

HARVESTING is the process of detaching a produce from the mother plant at the proper stage of maturity by an appropriate technique and as rapidly as possible with minimum damage or loss imparted to the commodity all at a relatively low cost. When quality is the primary consideration, not the cost, not the speed, and not the time, hand harvesting is the best approach. Generally, fruits intended for fresh market are hand harvested.

HAND HARVESTING

Hand harvesting has several advantages, especially with reference to selective picking at the right stage of maturity to allow for the maximum quality development in the fruit or vegetable prior to harvest (especially if the maturity stage is visually and easily assessible), with minimum damage done to the produce during picking and transfer to the appropriate container for subsequent handling, storage and transportation. In addition, it involves minimum or almost no capital investment and the harvest output is proportional to the number of people at work. The only capital required is perhaps shelter and transporation for the workers and small harvesting tools to separate the produce from the mother plant. One can supply more or fewer people based on needs.

The main problem with hand harvesting centers around the workers needed to do the job. There could be an acute shortage of labor when the need is the greatest during the harvesting season. Labor strikes in harvest time could occur and be very costly. The seasonal nature of fruit and vegetable crops makes it difficult to keep the entire work force engaged throughout the year. It would require very thorough planning and very efficient labor management. Various labor management approaches have been attempted, each with limited successes. Co-operative approaches employing the labor force for a variety of rotating jobs at locations within a reasonable distance have been successful in some sectors.

These are not just current problems; they existed in the past, they are there today and likely will continue to be there tomorrow. So is hand harvesting. Quality of fruits and vegetables is very important for successful marketing. It can only be satisfactorily obtained by hand harvesting at the right stage. Hence, even today, most of the fruits and vegetables intended for fresh market are almost entirely hand harvested.

MECHANIZED HARVESTING

Mechanical harvesting is employed for the majority of fruits and vegetables intended for processing where they normally are converted to other forms, where physical appearance is not a major consideration, where the commodity is consumed within a short time after harvest so the quality may not deteriorate seriously in spite of some damage induced by the mechanical harvesting techniques, and where the commodity is needed in large quantities.

Some advantages of mechanized harvesting are:

1. *Speed of harvest:* Substantial increase in the harvest output is possible with minimal trained labor. One mechanical harvester can potentially do the job of hundreds of manual laborers.
2. *Improved conditions for the workers:* Since only a few workers are needed to handle the mechanical harvesting equipment, better salaries, working conditions and other facilities can be provided, and the employees can be hired full time throughout the year for various activities. The workers, however, need the basic skills of operating the harvesting and other farm equipment.
3. *Reduced labor-related problems:* This is mainly because of the involvement of smaller numbers of people required, although strikes by these skilled workers could be equally costly.

The disadvantages and problems associated with mechanical harvesting systems are summarized below:

1. *Physical/mechanical damage to the crop:* Damage is one of the most serious problems associated with mechanical harvesting. Significant damage can be imparted to the commodity by the picking, screening and transferring devices or induced by the fall of the detached fruit on the lower branches of the tree or catching device. These damages serve as active sites for the invasion of pathogens, resulting in quick degeneration of the damaged crop, which can spread to surrounding crops if these are not effectively removed prior to storage and transportation.
2. *Non-selectivity:* The mechanical device cannot discriminate the fruit or vegetable based on maturity or color. Some selectivity may be possible with reference to size and shape. Grading of mechanically harvested crop is therefore a must and both size and color grading techniques are practiced. With hand harvesting it may be possible to do harvesting, grading and consumer packaging all at the same time in the field. For example, strawberries, blueberries, cherry tomatoes, lettuce, cabbage, etc. are picked and packed into the final containers in the field itself followed by quick cooling and transportation.
3. *Separation of plant debris:* The harvester generally picks up plant parts and foreign materials. These need to be sifted out, which is done by machinery in the field, while the rest is done in the packing house.
4. *Damage to fruit trees during harvesting:* Many fruit crops are mechanically harvested by shaking of the main trunk or branches and collecting the fallen fruits on a catch frame with a canvas platform. Especially the trunk shakers involve heavy duty pneumatic shaking operations and can cause considerable damage to the tree if operators are not properly trained.
5. *Large volume output:* The large volume output is an advantage in most occasions and is one of the reasons for employing mechanized harvesting, but this could become a limitation in the absence of adequate processing and handling capacity.
6. *Expensive machinery:* Machinery often needs a large capital investment. Harvester designs are very specific for a certain type of produce based on its physical characteristics and growing conditions. These often can change due to altered farm practices, use of high yielding varieties, etc. Such changes may require redefining the selec-

tion criteria for harvesters. Hence, what is best today may quickly become obsolete even before the equipment gives the expected returns.
7. *Social impact:* One machine will likely do the job done by hundreds of workers. So if deployed on community farms where the majority of people thrive on farm labor, machines could cause labor unrest.

The ever-increasing cost of manual labor, decreasing availability of timely labor and the need for high-speed high-volume output probably provide the best incentives for the development of mechanical harvesting equipment in spite of the various problems and disadvantages associated with them.

ELEMENTS OF HARVESTING: HAND/MECHANICAL

- *Detection:* This is possible only with hand harvesting operation.
- *Selection:* This also is possible only with a hand harvesting operation.
- *Detachment: Mechanical devices* involve combing, raking, pulling, lifting, digging, cutting, shaking, blowing, sucking and other operations. *Hand held devices* feature a knife, clippers, picker pole with a knife, bag.
- *Collection:* Mechanical harvesters like trunk shakers require a catch frame with canvas; others require a conveying device to fill large bins. Sometimes these are part of the harvester and most often after some preliminary cleaning, the harvesters feed a receiver truck that follows the harvester. For hand harvesting, this will initially involve small baskets or bags emptied into a storage bin.
- *Separation:* This involes the cleaning operation to separate unwanted material. Usually, this is done in the field, and the rejects are left in field.
- *Handling:* From harvester to transport trucks to packing house.

IMPORTANT FIELD/ORCHARD-CONSIDERATIONS FOR MECHANIZED HARVEST

1. *Genetic considerations:* The selected variety should be suitable for mechanical harvesting. The tree, for example, should have the strength to withstand the mechanical shaking operation. The crop should preferably attain a uniform maturity by harvest time, facilitating once-over harvesting. The crop also should have the mechanical strength to withstand aggressive mechanical handling.

2. *Planting system and tree training:* High-density planting of vegetable crops is ideal for once-over mechanical harvesting needs. High-density planting also is recommended for tree fruits because it favors dwarf trees. Tree fruits which are separated by shaking or blowing devices should be trained to take high-frequency pneumatic shaking. Dwarf trees are preferable to tall trees and similarly short branches are preferred over long branches to facilitate the effective transfer of the vibration induced by the trunk or limb shakers to the fruits and nuts. The longer the tree or branch, the less effective is the fruit separation.

3. *Crop control—Pruning:* The main reason for pruning is to maximize the fruit yield and value—it is a balance between fruiting and fruit growth; in other words, it is a balance between the number and size. Instead of having a large number of smaller fruits which are labor intensive for harvesting and handling, the trees are pruned to remove some growing parts and nonproductive branches. The process helps to produce larger fruits, although in relatively smaller numbers. Pruning effectively reduces the number of growing points and improves the vigor of those remaining, thereby increasing the crop potential. A heavy bloom and fruitset likely damage the plant and normally result in weak vegetative growth. If much of the excess can be removed before the bloom, the potential for the vegetative vigor to size the new crop and to produce healthy buds for a crop the following year is maximized. These are desirable from a mechanical harvesting point of view since they increase the strength of the mother plant.

    Fruit thinning is another concept whereby one facilitates the development of a smaller number of large fruits rather than numerous smaller ones. In this case the excess flower buds are removed either by hand or by the use of chemicals. Thinning involves: (1) application of a caustic spray during the period of bloom, (2) application of growth regulators after fruit set, or (3) mechanical or hand thinning.

4. *Harvest Control:* A certain amount of flexibility can be achieved for harvesting of some fruit crops by additional orchard management techniques. Chemicals applied in the field can either hasten the ripening of fruits or delay the ripening. Ideally, by spraying of chemicals that hasten the ripening stage to a section of the orchard the harvest date can be advanced. Similarly, another section of the

orchard can be sprayed with a chemical that delays ripening so this section could be harvested at a later date. Untreated areas could be harvested in between, thereby effectively prolonging the harvest season. Growth regulators like Alar act as growth inhibitors on some fruits and a growth promoter on others. Alar retards, for example, ripening of pome fruits and accelerates ripening of stone fruits. It also intensifies the red color of some varieties of apples. However, recently there have been several adverse reports on the use of Alar on apple orchards. Ethephon, an ethylene releasing agent has been shown to advance the maturity date. To control the fruit drop, the following chemicals have been used: 2,4,5 T (2,4,5 trichloro phenoxy acetic acid), NAA (naphthalein acetic acid), CPA (p-chloro phenoxy acetic acid) and alar (applied 60–70 days prior to harvest—apples).

TYPES OF MECHANICAL HARVESTERS

Various types of mechanical harvesting machines have been developed, mostly with each being specific to a particular commodity. These can be broadly grouped as:

1. Direct contact devices
2. Vibratory devices

Direct contact devices are based on actions such as cutting, pulling, snapping, twisting, stripping, digging, lifting, etc. Most vegetable harvesters with some used for non-tree fruits come under this group.

The vibratory devices are those which use trunk and limb shakers with a catch frame and canvas. These have been widely used for apples, apricots, cherries and nuts.

HARVEST AIDS

Harvest aids are mechanical devices and supports that assist in improving hand harvest operations. They include:

1. Picker poles with a knife and a canvas bag, for many tree fruits
2. Multi or single station platforms to properly position workers, used with tree fruits.
3. Conveyor belts to move harvested heavy fruits and vegetables, such as pineapples and melons.

4. Harvesting cages which are similar to platforms
5. Lights for night harvesting

EXTENT OF MECHANIZATION

Kader (1985) classified the extent of vegetables and fruits that have been harvested by devices, hand or mechanical. The percentages which follow are in terms of extent of mechanical harvesting. What is not mechanically harvested is hand harvested. Those with an asterisk are commodities more than 50% of which are processed.

Vegetables

- *Mechanically harvested 0–25%:* Artichoke, asparagus, broccoli*, cabbage, cantaloupe, cauliflower, cucumber*, lettuce, green onion, cress, dandelion, eggplant, endive, fennel, kale, kohlrabi, mushroom*, okra, pepper, rhubarb*, Romaine, squash, watercress, cassava, ginger, parsley root, parsnip, rutabaga, turnip, taro.
- *Mechanically harvested 25–50%:* Sweet potato, mustard greens, parsley, Swiss chard, turnip greens
- *Mechanically harvested 50–75%:* Snap beans*, dry onion, pumpkin*, tomato*.
- *Mechanically harvested 75–100%:* Carrot, potato*, lima bean*, snap bean*, sweet corn*, peas*, spinach*, horseradish*, beet*, garlic, Brussels sprouts*, radish.

Fruits

- *Mechanically harvested 0–25%:* Apple, apricot, avocado, banana, sweet cherry*, grape*, guava*, kiwi, kumquat*, lychee, mango, nectarine, peach*, pear*, persimmon, pineapple*, pomegranate, wild blueberry*, currant, gooseberry, strawberry, grapefruit, lemon*, lime, orange*, olive*, papaya, passion fruit*, tangerine, cashew, coconut*, chestnut*.
- *Mechanically harvested 25–50%:* Red raspberry, macadamia.
- *Mechanically harvested 50–75%:* Prune*, blackberry*, highbush blackberry*, black raspberry*, pecan.
- *Mechanically harvested 75–100%:* Tart cherry*, date, fig, cranberry*, almond, peanut*, pistachio, walnut.

The reader is referred to Ryall and Lipton (1978) and Ryall and Pentzer (1979) for more detailed discussion on the harvesting of individual fruit and vegetable crops. The following figures illustrate some harvesting techniques and harvest aids used in the industry.

FIGURE 5.1. Tree shaker and illustrated shaking operations.

A canopy shaker in operation

FIGURE 5.2. Canopy shaker.

Harvesting of Corn

FIGURE 5.3. Corn harvesting illustrations.

Cucumber harvesting

Peppers

FIGURE 5.4. Cucumber and pepper harvesters.

Carrot and Potato Diggers

FIGURE 5.5. Diggers for carrots and potatoes.

Onion Harvesters

Pull Behind Onion Harvester

planter

FIGURE 5.6. Onion harvester.

Sickle and Cutter
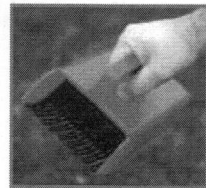
Rake/scoop for cranberries
Cutter for asparagus

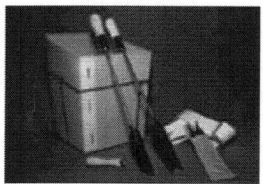

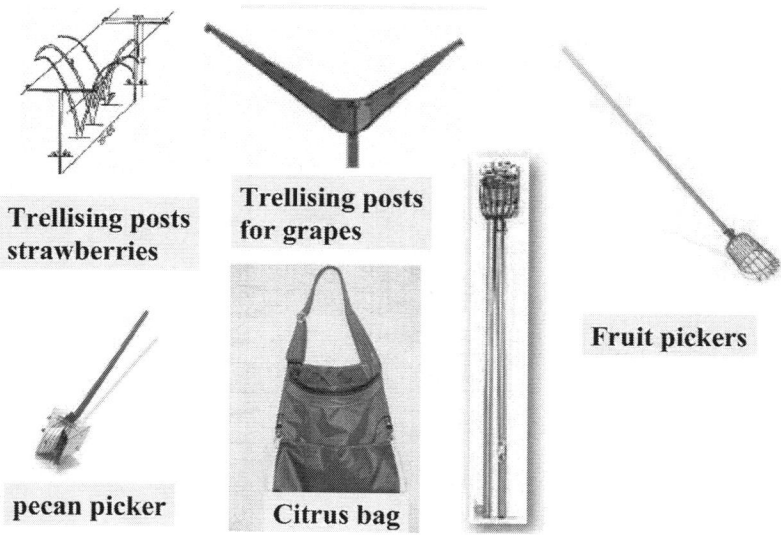

FIGURE 5.7. Harvest aids.

CHAPTER 6
Quality and Maturity Indices

INTRODUCTION

QUALITY, to be meaningful, can only be defined with reference to a particular context, and in this case a specific commodity intended for a specific use. In reality, it has be defined clearly to indicate what we want it to represent. An excellent quality for one purpose may represent a poor quality for the other. For example, a ripe fruit with fully developed flavor and color may have excellent edible or eating qualities but would make a very poor choice if the fruit has to be stored or transported for later use. The quality of vegetables suitable for canning purposes may be too mature for preservation by freezing or for consumption in the fresh state. In this section, we address the quality issues of fruits and vegetables with reference to their suitability for the purpose of storing them for as long a period as feasible prior to consumption, and with respect to their sensory and nutritional qualities. In this context, quality of fruits and vegetables is a combination of various physico-chemical and nutritional characteristics that give value or wholesomeness as a food item for the consumer. Obviously, quality is a multi-component system and generally includes "appearance or visual/esthetic factors, texture or mouth/finger feel factors, flavor (taste and smell), nutritive value and safety/antinutritional factors" (Kader *et al.*, 1985). The relative importance of each quality factor depends upon the commodity and its intended use (fresh or processed).

The appearance factors play a dominant role in the selection of the commodity for various purposes. This includes the physical character-

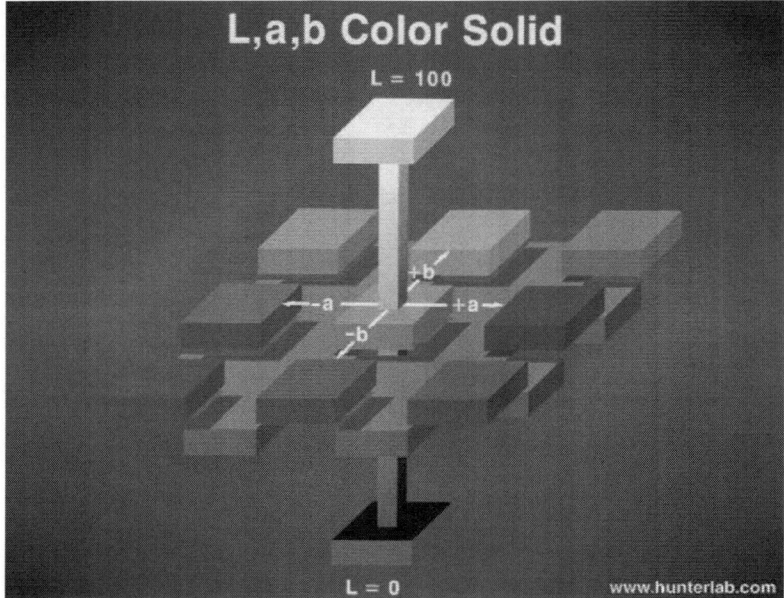

FIGURE 6.1. Hunter L, a, b, model.

istics such as size (dimensions, weight, volume), as well as shape and form (diameter/depth ratio, smoothness, compactness, uniformity). It also includes the color and gloss that are attractive to the eye. There are instrumental methods to quantify color, which are based on the fact that color can be expressed as mathematical combination of three primary colors (red, blue and green). Tristimulus colorimeters, the most commonly used instruments to objectively measure color of foods, make use of different models to express color values, but the most common and widely accepted one is Hunter's L, a, b (Figure 6.1). L value depicts the lightness of food particles, and it is expressed on a scale of 0 (dark) to 100 (light). Similarly, a, and b values are expressed on the scale of –60 to +60, where a value indicates degree of greenness/redness and b value indicates degree of blueness/yellowness.

As far as appearance is concerned, defects can be numerous and internal or external. Examples of morphological defects include sprouting of potatoes, onions and garlic, elongation and curvature of asparagus, seed germination in tomatoes, peppers, etc. Physical defects include shriveling and wilting; mechanical damages such as punctures, cuts, crushing and abrasion are possible. Physiological defects include chilling and freezing injury, internal breakdown in fruits, water core in apples, black heart of potatoes, etc. Pathological defects include decay caused by fungal and bacterial activities: blue mold, grey mold, etc. Entomological defects include damages caused by insects, pests, etc.

The textural characteristics basically are the "feel" properties, whether finger feel in terms of firmness, hardness, crispiness, etc. or mouthfeel in terms of succulence, juiciness, tenderness, mealiness or grittiness. It can include many other terms such as cohesiveness, chewiness, brittleness, stickiness, etc. Textural characteristics are commonly studied in food industries and also in food-related researches by using objective methods of measuring textural parameters. The instrumental texture profile analysis method has become very common with the advent of computer-assisted texturometers. The texture analyzers generate a "texture profile" by compressing bite-sized food particles to give force vs time curves (Figure 6.2). The compression test in a texturemeter basically simulates the mastication (chewing) process of humans.

Several instruments can be used to measure texture profile analysis. Some are illustrated in Figure 6.3.

Flavor is a combination of taste and smell. Only a taste panel can give a meaningful interpretation of flavor characteristics and it is a highly subjective evaluation technique. The chromatographic techniques, such as gas chromatography, can be used to objectively evaluate flavor components. Objective techniques are fairly limited in terms of providing the full picture of a flavor profile. But selected components can be monitored to follow the changes in the flavor profile if correlated well with a trained taste panel.

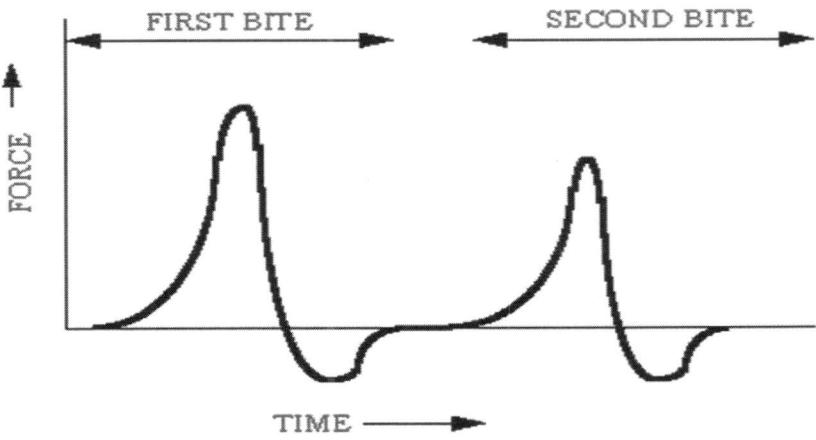

FIGURE 6.2. A typical texture profile.

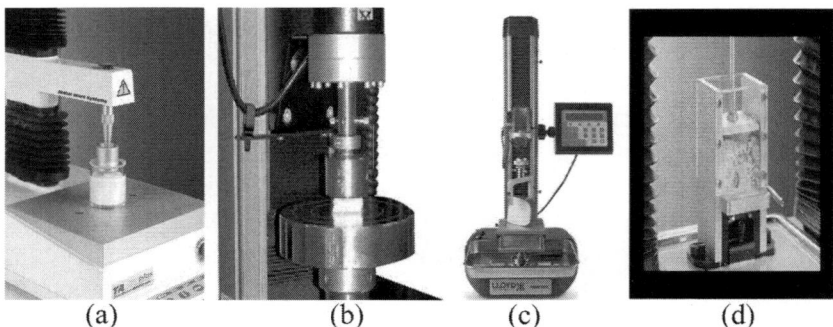

FIGURE 6.3. *Some instruments used for texture profile analysis. (a)Texture Technologies (b) Instron (c) Lloyd Instruments (d) Brookfield.*

Nutritional and antinutritional characteristics were discussed, with some fruits and vegetables as sources of several nutrients, especially the vitamins and minerals. Fruits and vegetables can contribute antinutrional elements as well. Contaminants such as chemical residues (left over from field applications or those absorbed by the system), heavy metals, etc., are monitored routinely by various agencies (like Agriculture Canada in Canada, U.S. Department of Agriculture, Food & Drug Administration in USA) to assure compliance with established maximum tolerance levels.

In moving from the field to the consumer, produce passes through several middlemen, each intervening with the intention of earning a profit for his/her efforts. People concerned are the grower or the producer, the shipper, the receiver, the distributor, the retailer and the consumer. Quality perceptions differ for people in each category. Appearance factors in terms of good color, uniform size, shape and freedom from defects are considered the most important aspects of quality by almost everyone concerned from grower to the consumer. In addition, the grower who has to produce the commodity in the field considers that the commodity also must score high on yield, disease resistance, ease of harvest and shipping quality. The plant breeder traditionally has given these characteristics a higher priority over nutritional and flavor quality (Kader *et al.*, 1985). To the shipper, receiver, distributor and retailer, firmness/texture and long storage life are the primary concerns, in addition to the appearance factors. For the most important person, the consumer, high quality in fruits and vegetables includes attractive appearance, firmness to the touch and good flavor characteristics as well as nutritive value. Although the consumer normally buys on the basis of appearance and feel, the satisfaction is dependent on good eating characteristics. If quality is not satisfactory upon consumption, the consumer is not likely to buy the same kind from the same source again.

These various factors generally are used to identify the suitability of fruits and vegetables for the intended use. They also are used to lay down specifications to meet different standards. The grades and standards in force in different countries are given in a table at the end of the chapter. The table provides a comprehensive coverage of the standards for almost all fruits and vegetables and should serve as a good reference material. Canadian standards follow very closely the U.S. standards. These guidelines and standards are largely voluntary, except when required by state and local regulations, by industry marketing orders or for export marketing. They also are used by many private and government procurement agencies when purchasing fresh fruits and vegetables.

FACTORS INFLUENCING QUALITY

So far we have only looked at quality attributes. With reference to fruits and vegetables, we know there are numerous factors that influence the quality attributes. These can be discussed under the following headings:

Genetic Factors

Genetic factors include selection of cultivars, rootstocks, etc., which have a tremendous bearing on the produce quality.

Environmental Factors

Environmental factors include both climatic factors (temperature, light, rainfall, irrigation, wind) and cultural factors (soil type, mineral nutrition, water supply, pruning, thinning, pesticides, growth regulators).

Temperature

All plants have a tolerable growth temperature range. As a general rule, the higher the temperature within this tolerance range, the faster the growth. Again, higher temperatures within this range also are likely to result in an earlier harvest. This factor has been used as a basis for the temperature summation technique for determining optimum harvest maturity. This is based on the premise that everything else being the same, the time a fruit takes to fully develop is dependent on the surrounding temperature. Warmer temperatures permit faster development because they are physiologically more favorable for growth. The temperature or heat accumulation process is carried out mainly with reference to a base or reference temperature only above which there can be any significant growth. The base conditions vary depending on the commodity. The base conditions and accumulated heat units required for maturity/harvest are given in Table 6.1.

It is expressed in degree-hours or degree-days. A degree-hour is the accumulated heat unit equivalent to the exposure of the crop to one degree above the reference temperature for one hour. A degree-day is similarly equivalent to a similar exposure for one day or 24 hours. A degree day = 24 degree hours.

Example: For peas the base temperature is 40°F. On a day when the temperature is 60°F, each hour constitutes 60 − 40 = 20 degree-hours. A full day with an average temperature of 55°F, is equivalent to 15 degree-days. It takes approximately 1200 degree-days for peas from planting to reach harvest maturity. At 55°F average temperature conditions, it would, therefore, take about 1200/(55 − 40) = 80 days from planting to harvest.

General formula:

Mean heat units = (Actual temperature − Reference temperature) × time

TABLE 6.1. Accumulated Heat Units for Harvest Maturity.

Commodity	Base Temp. (°F)	Mean Heat Units (deg-day)	From	To
Peas	40	1200–1700	Planting	Opt. maturity
Corn	50	2000–2200	Planting	Opt. maturity
Asparagus	40	440–620	Planting	1st cut
Snap bean	50	1150	Planting	10% seed
Tomato	55	1350	Blossom	Ripe
Lettuce	40	1400–1700	Planting	Harvest
Apples	40	4400–5000	Blossom	Harvest ripe
Cherry	55	950	Blossom	Ripe

One must be careful in extrapolating the results of temperature summation. For example: sweet peas are a temperate crop and are best grown in hilly regions. The reference or base temperature for pea growth is 40°F (below which they can hardly grow). The optimal temperature is about 60°F. Accumulated heat units for maturity is ~1200 degree-days. One might say it would take about 60 days from planting to harvest under optimal temperature conditions. However, temperatures higher than the optimum (60°F), may not produce the same result and can cause heat damage to the plant or retard the growth and thereby decrease the productivity.

From a climatic point of view, warm days and cool nights are ideal for plant growth. Such acombination is possible in temperate climates. The higher temperatures during the day permit greater photosynthetic activity and therefore better growth and accumulation of food reserves. The cool nights will keep the produce respiration rates at a lower level and hence result in lesser depletion of the stored reserve. The same commodity grown under tropical conditions may not yield produce of the same quality; however, it is likely to be sweeter. In tropical climates, the night temperatures usually are not very different from the day temperatures and hence the respiratory activity at night also is higher, which might deplete some fruit quality. For the tropical fruits, tropical conditions will yield better quality fruits than temperate conditions, in which they still may be able to grow, but rather more slowly.

Light

The quality of produce is influenced by the following characteristics of light: duration, intensity and quality.

- *Duration:* The length of the day/night is important. Longer days and shorter nights are better than converse (more photosynthetic than respiratory activity)—the results are reflected in the composition of the fruit/vegetable (sugars, starch content, etc.).
- *Intensity:* The higher the intensity, the sweeter and less acidic the fruit will be (examples: citrus, banana, mango). Also, higher intensity may result in greater fading of some colors. Canopy shading is employed for some products: tomatoes show deeper red color; cucumber show deeper green color under shaded growth conditions. Density of planting also will influence the intensity of light; for example, under high-density planting, fruits are likely to be less sweet.

- *Quality:* The quality of light also is important in the development of pigmentation in fruits and vegetables. Purple cabbages or eggplants derive their color from exposure to light in the shorter wavelength (blue and violet); direct exposure to sunlight will result in green color.

Other environmental factors include rainfall, soil texture, wind factors, etc.; all play a role depending on the type of produce.

Cultural Factors

Mineral Nutrition

Materials were discussed in Chapter 4.

Other cultural practices: Pruning, fruit thinning, planting: high- vs low-density were discussed in Chapter 5.

Chemical Sprays/Growth Regulators

Chemical sprays and growth regulators in fields and orchards are numerous; they are applied in various stages and forms. Their influence on postharvest quality of the produce has not been clearly understood in most cases.

Some examples:

Pesticides: Petroleum oil sprays used to control citrus pests have been shown to result in undesirable effects on quality (low TSS and high acid).

Abscission-preventing chemicals are used to control the fruit drop: 2,4,5 T (2,4,5 trichloro phenoxy acetic acid); NAA (naphthalein acetic acid); CPA (p-chloro phenoxy acetic acid); Alar (applied 60-70 days prior to harvest of apples).

Growth hormones have been used for inducing fruit setting and for increasing the size of the fruit: Alar (prebloom application); Ethaphon (prebloom application); GA (gibberellic acid).

Chemicals used as sprout inhibitors: MH (Maleic hydrazide) commonly used on onions; MENA (methyl ester of NAA).

Fruit-thinning chemicals: NAA, CPA, Ethephon.

Growth regulators (altering the maturity date): Alar: acts as both growth inhibitor and growth promoter—retards ripening of pome fruits while it accelerates ripening of stone fruits. It intensifies the red color of some varieties of apples. Ethephon advances the maturity date.

Chemicals in postharvest applications: Numerous applications have been used: fungicides, fungistats, disinfectants, fumigants, sprout inhibitors.

Harvesting Stage

With so many factors influencing the growth and quality of produce, it is not surprising that the best stage to harvest a commodity is highly variable. It may only be possible to offer guidelines as to when to harvest with reference to a given commodity species at a specific location.

The harvesting stage is, however, the most important factor in the postharvest quality of any produce. It determines its storage life, eating qualities, market potential, yield, etc. A lot of things depend on it, and it is one of the prime concerns of the producer, processor and the consumer. In order to address this, let us look at answers for three main questions:

1. What parameters can we choose to determine the optimal stage of maturity?
2. How can we use these parameters to indicate the maturity of the crop?
3. How we can use these parameters to predict when the crop will be ready for harvest?

Generally speaking, harvest or maturity indexes are good indicators of quality as well. On the other hand, we can have many important quality indexes that are not good indicators of optimal maturity. First, let us be clear as to what we mean by optimum maturity. The consistent definition of optimum maturity appears to be "that stage of maturity at which a commodity has reached sufficient stage of development that after harvesting and postharvest handling (including ripening where required), the quality will be acceptable to the ultimate consumer" (Kader *et al.*, 1985). This type of definition is needed primarily because for several fruit crops such as banana, pineapple, etc., the optimal maturity at harvest will be far from when the crop attains the most desirable eating quality. With most vegetables, normally the optimal maturity coincides with optimal eating quality.

Horticultural maturity indicates when the crop possesses the prerequisites for use by the consumer for a given purpose. In other words, a commodity can be horticulturally mature at any stage of development. For example, sprouts are horticulturally mature at a very early stage, compared to the roots and tubers. In the order of horticultural maturity, we have sprouts, stems and leaves, inflorescence, fruit vegetables, fruits and underground storage organs and, finally, the seeds. Overall, this delineates development from start to finish. Growth ceases. The plant organ starts to mature and goes through ripening and senescence if left long enough on the plant.

Let us look at some of the changes that take place as a fruit or vegetable starts growing on the plant. Figure 6.4 illustrates some typical physico-chemical changes that occur during the ripening of tomato fruit. Chlorophyll content of the fruit is decreasing as the fruit turns from the mature green stage to the ripe red stage. At the same time, the lycopene content is increasing. In other words, the parameters chlorophyll and lycopene contents of the fruit correlate with the disappearance of the green color and appearance of the red color in the mature tomatoes. The CO_2 evolution rate is increasing up to the pink stage and then declining a little bit. Ethylene evolution increases continuously. CO_2 evolution, a

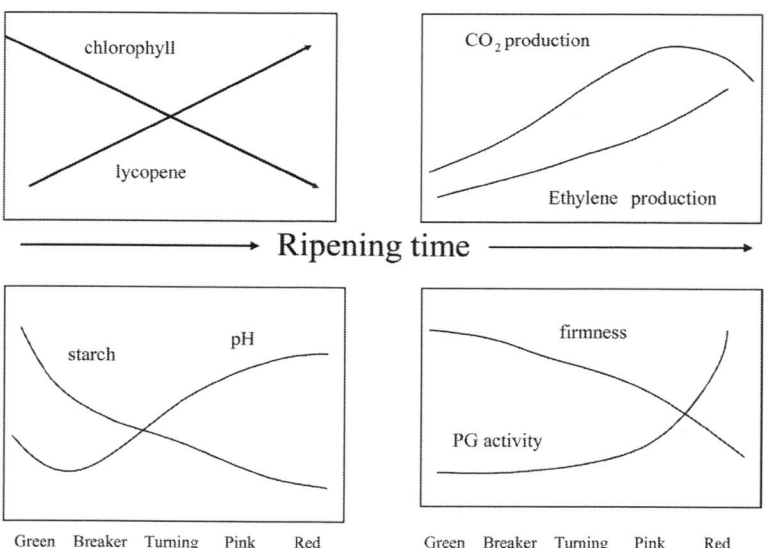

FIGURE 6.4. Typical changes taking place in tomato during ripening.

measure of respiratory activity, indicates the physiological state of the fruit. The starch content is decreasing, and the soluble solids content (sugars) is somewhat steady. The fruit would taste somewhat sweeter at the ripe stage, not necessarily because the sugar content is increasing but, in this case, because the starch is disappearing, thereby exposing the sugars. At the same time the pH is increasing, indicating the fruit is less acidic. Firmness and polygalacturonase activity are also correlated. As the fruit is ripening, the firmness is decreasing; basically, the pectic substances are degraded. Pectins are esters of polygalacturonic acid and are acted upon by the enzyme polygalacturonase. You will see a perfect correlation between the enzyme activity and the resulting action.

Another example is the various physico-chemical changes during the growth of French beans as illustrated in Ramaswamy et al. (1981). Physical parameters like length, weight, and perimeter; chemical parameters like ascorbic acid, alcohol, insoluble solids, sugars, and crude fiber are related to harvesting age of the vegetable. The activity of the enzyme peroxidase and a texture parameter, shear force, which is the force required to shear a bean pod, also are included. Several of these parameters change continuously as the bean matures. From a sensory quality point of view, it was reported in this study that the optimal stage of maturity was between 12 and 16 days. While some parameters like length, weight, perimeter, shear force, etc., change in a fairly linear fashion, others change dramatically as they pass the stage of optimum maturity. While these other parameters are good quality indexes that can differentiate tender beans from the mature ones, they are not very meaningful in terms of maturity indexes. This is because they remain relatively constant until the optimal stage of maturity and then start increasing. How can you expect to employ such a parameter to predict the harvesting stage? On the other hand, parameters like length and weight can be consistently correlated with the maturity days with reference to the particular cultivar over several seasons in comparable growing areas; their relative value could indicate when the crop will be ready for picking. For example, let us say the optimal length is 14 cm in beans that possess good picking qualities. The length increases from about 10 cm to 18 cm in 12 days (from 8 to 20 days). In other words, the rate of increase in length is 8/12 or 0.66 cm per day. On a particular day if you find the length is 6 cm, then to reach 14 cm would require 8/0.66 = 12 more days.

Measurement/prediction of the stage of optimum maturity for harvesting sounds like a simple technique when we treat the data in the above manner. However, it is much more complicated. We looked at various factors that influence quality. In fact, they all influence the growing pattern of the plant as well. In addition, on the same plant you may have organs of varying maturity. With hand harvesting, you may set certain criteria for picking, say in the above example you might say not to pick beans smaller than 12 cm. In the case of tomatoes, you could recommend pick ones that are just breakers or turning pinkish etc., depending on the end use. With mechanical harvesting, one has to make an intelligent guess as to when the bulk of the field will be at the optimal stage.

After considerable research in this area, various parameters have been selected to be good indicators of harvest maturity. The meaning and nature of several of the factors that are listed are to an extent self-explanatory. We shall briefly go through a few.

Elapsed days from full bloom to harvest: This has been used for apples and pears with some success. Basically, given the appropriate nutrients, water and sunshine, these fruits will mature a given number of days after flowering. By keeping accurate data over several seasons, one might guess with some reasonable assurance the probable week when the orchard will be ready for harvest.

Mean heat units: This has been used much more widely than the elapsed days from bloom. This is basically based on the fact that everything else being equal, the time a fruit takes to fully develop is dependent on the surrounding temperature. The warmer the temperature, the faster the development within a short range of temperatures which are physiologically favorable for growth. The range mainly refers to a base reference temperature, which varies from commodity to commodity.

Specific gravity: Density or specific gravity grading is sometimes employed to separate produce based on maturity. As the produce matures, its specific gravity generally increases. Practical grading of processed peas using brine flotation technique is a standard practice. Brine (salt) solution of different concentrations are selected; normally these are used in a series. Shelled peas are dipped in the most dilute solution first. The floaters in this treatment generally are rejected since they mostly comprise defective peas, pods or leaves, etc. The sinkers then go to the next concentration; here the floaters will be the tender peas: Grade A. Sinkers go to the next floaters, which belong to the grade B. The sinkers go to the next solution. Floaters from this will be Grade C and the sinkers are rejected as they are too mature. During this process, the shelled peas also will absorb some salt—not a serious problem because these are normally canned in a brine solution. This technique has also been used for cherries and watermelon.

Chemical Methods: Most of the parameters listed under this category have standardized methods. They are listed in the Official Methods of Analysis of AOAC. The remaining are parameters related to color and texture of the produce. Both have been used in recent years. Color as a criterion for quality grading and evaluation is even more interesting, since it can be applied without destroying a portion of the produce.

CHAPTER 7

Post Harvest Physiology of Fruits and Vegetables: Respiration

INTRODUCTION

FRUITS and vegetables are living tissues and they should be alive in order to maintain their keeping qualities. Like all living entities, they respire, consuming oxygen and liberating carbon dioxide. The principle task of the respiratory process is production of energy needed for various metabolic activities. They rely on organic matter that generally is replenished during growth. Upon removal from the mother plant, the fruits and vegetables are cut off from their normal supplies of water, minerals and organic matter, provided to them from other parts of the plant. The produce, however, remains capable of continuing a wide range of metabolic activities, breaking down stored organic matter in order to meet energy requirements. Some of these physiological activities are highly desirable for the attainment of the optimum eating qualities: example, ripening of fruits like banana, mango, papaya, pineapple, etc. These activities also are desirable because they help to maintain the vigor of the tissue and provide a kind of defense against attack by spoilage organisms. Otherwise, they are less desirable and unnecessarily result in loss of stored organic matter. But, however undesirable they are, they still have to be continued at some minimal level in order to maintain quality.

The primary physiological activity in fruit and vegetable tissues is respiration, which is the complex process of oxidation of organic matter (starch, sugars, acids, fats, proteins, etc.) to simpler molecules, such as CO_2 and H_2O with the concurrent production of energy and other intermediate products.

The undesirable consequences of respiration in produce are:

1. Loss of food value (stored organic matter is degraded).
2. Hastening of senescence (process of aging).
3. Loss of saleable weight.
4. Reduced quality (usually after a ripening process for fruits).

On the other hand, respiratory activity has several desirable functions:

1. It provides the energy for numerous metabolic processes.
2. It provides valuable intermediates.
3. It maintains tissue vigor.

In summary therefore, respiratory activity is both good and bad. It is needed to sustain life; but a high respiratory activity results in rapid burning up of food reserves and a loss in product quality. Our objective, while considering a long storage life for the commodity is to let the fruits and vegetables respire to maintain their life-sustaining activities, but to maintain them at the minimum possible levels to reduce the rate of quality loss. In order to do so, we must have an understanding of the respiratory process, the various factors that influence respiratory activity and which of these we can practically alter during storage to maintain a low respiratory activity.

AEROBIC vs ANAEROBIC RESPIRATION

Respiration can take place with or without oxygen. Respiration in the presence of a vast excess of oxygen is called *aerobic respiration,* while respiration in the absence or limited availability of oxygen is called *anaerobic respiration.* Both *aerobic* and *anaerobic* respiration yield energy; however, anaerobic respiration is undesirable in fruits and vegetables because it results in off-flavor development due to formation of fermentation products like ethanol and acetaldehyde. Further, the highest yield of energy comes from the aerobic respiration.

ENERGY CURRENCY—ATP

The reactions that take place in a plant tissue are complex and numerous. Both energy-yielding and energy-absorbing activities occur in the same tissue. On one side, the simple sugars may be utilized by the respiratory process to derive energy needed for the synthesis of other vital macromolecules, such as the more complex carbohydrates, proteins or fats. The formed fat, protein or carbohydrate can then be degraded as needed to provide additional energy for other activity. The chemical reactions and energy transfers that take place in plants are very efficient processes, unlike manmade processes, such as internal combustion engines, electric generators, etc., which also involve transformation of energy from one form to the other. Plant cells are provided with an energy-trapping device in the form of ATP—adenosine triphosphate—which is commonly termed as *the energy currency* of the plant. Whenever energy is released by the oxidation of carbohydrates, lipids or proteins, it is immediately utilized in the synthesis of ATP from ADP (adenosine diphosphate) and inorganic phosphate. And when energy is needed, ATP is downgraded to ADP and inorganic phosphate. ATP, however, is not the *energy-storing apparatus* of plant cells. The energy-releasing and energy-conserving activities are represented in Figure 7.1.

AEROBIC OXIDATION OF GLUCOSE

In simple terms, the aerobic oxidation of glucose can be represented by its conversion to carbon dioxide and water with the liberation of heat (exothermic reaction). During aerobic oxidation, glucose combines with oxygen to produce carbon dioxide, water and heat, as represented in the equation:

$$C_6H_{12}O_6 + 6O_2 = 6CO_2 + 6H_2O + \text{Heat}$$

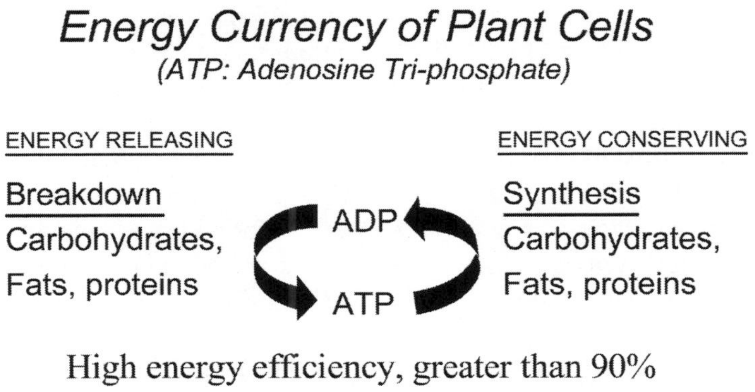

Energy Currency of Plant Cells
(ATP: Adenosine Tri-phosphate)

ENERGY RELEASING	ENERGY CONSERVING
Breakdown	**Synthesis**
Carbohydrates, Fats, proteins	Carbohydrates, Fats, proteins

ADP ⇌ ATP

High energy efficiency, greater than 90%

FIGURE 7.1. *Energy releasing and conserving reactions.*

Balancing the equation, one can find that a single molecule of glucose needs six molecules of oxygen to produce six molecules of carbon dioxide and six molecules of water. Calorimetric determinations have shown that the amount of heat energy liberated per mole of glucose is about 673 kcal. So we can rewrite the equation as:

$$180g\ C_6H_{12}O_6 + 192g\ O_2 = 264g\ CO_2 + 108g\ H_2O + 673\ kcal\ Heat$$

Respiration rate generally is expressed in terms of oxygen consumed, carbon dioxide liberated or heat liberated per unit weight of the produce in a given time:

O_2 *basis:* mg or mL of O_2 consumed/kg(produce)/h

CO_2 *basis:* mg or mL CO_2 produced/kg(produce)/h

Heat basis: kcal/ton(produce)/day

Using the above equation as the basis, one unit can easily be converted to the other.

1 mg O_2/kg-h = (264/192) mg CO_2/kg-h or 1.375 mg CO_2/kg-h

or

1 mg CO_2/kg-h = (192/264) mg O_2/kg-h or 0.0.727 mg O_2/kg-h

or

1 mg CO_2/kg-h = (673/264000) kcal/kg-h or 0.00255 kcal/kg-h
= 0.00255 × 1000 × 24 = 61.2 kcal/ton-day

It has been observed that although 673 kcal of energy is liberated when one mole of glucose is oxidized, not all of this generally is released in the form of heat. Some of it is used as chemical energy to support the various biochemical reactions taking place in the plant system. Especially when the organ is attached to the mother plant, a vast amount of energy is used in the form of *chemical energy* with the formation of ATPs from ADPs as sugars are oxidized. Then the ATPs are expended to form ADPs as the liberated energy is utilized for the synthesis of various storage reserves. Before harvesting, although synthetic reactions continue with the constant demand for vital components (such as synthesis of cell wall and cytoplasmic materials), a major amount of energy is generally lost as heat. In most postharvest design applications, 673 kcal/mole is used as the basis in computing refrigeration requirements.

BIOCHEMICAL PATHWAYS

The earlier equation representing glucose oxidation is only a simplified overall reaction. The biochemical reactions go through a sequence of metabolic pathways trapping the liberated energy as ATPs in several stages. Oxidation of glucose follows two major biochemical pathways: (1) the *EMP pathway* or *glycolysis* and (2) the *TCA* (tri-carboxylic acid), *Krebs* or *citric acid* cycle.

Glycolysis or EMP Pathway

Glycolysis refers to the conversion of glucose to pyruvate. EMP stands for Embden-Meyerhof-Parnas. The series of reactions that take place resulting in the formation of pyruvic acid are shown in Figure 7.2. The enzymes in the initiation and propagation of glycolysis are known as "kinases." These enzymes catalyze phosphorylation reactions. The first step is the conversion of glucose to glucose-6-phosphate, a phosphorylation reaction catalyzed by the enzyme *hexo-kinase* meaning it is the *kinase* enzyme acting on a hexose, glucose in the above example, leading to glucose-6-phosphate which exists in isomerism with fructose-6-phosphate.

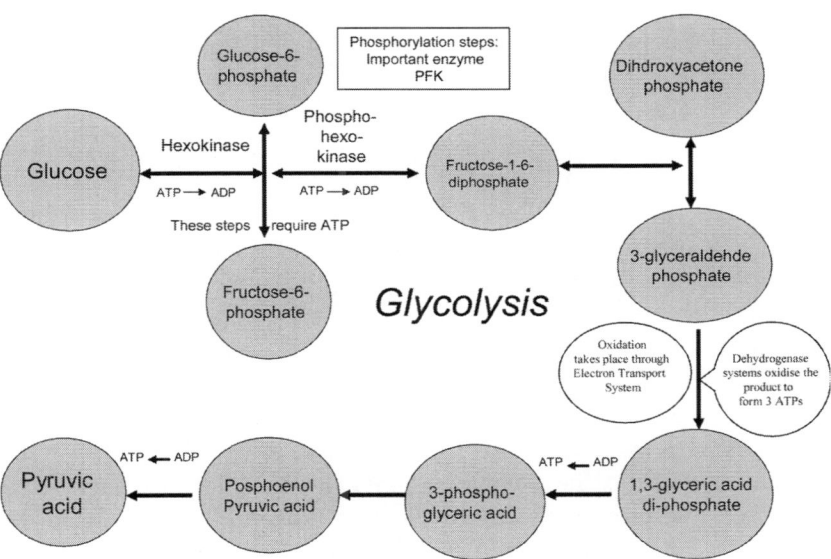

FIGURE 7.2. EMP-pathway (glycolysis).

The next step is the further phosphorylation to fructose-1-6-diphosphate by the enzyme *phospho-fructo-kinase* or PFK in short. This is a *key enzyme* in the glycolytic process. Plant cells control their energy supply by controlling the activity of this enzyme. Enzyme PFK is inhibited by the presence of excess ATP, although it actually needs energy from ATP for some phosphorylation reactions. What follows is a series of reactions resulting in splitting of the six-carbon glucose to three-carbon glyceraldehyde, which goes through transformations and dephosphorilation leading to the formation of pyruvic acid. Following the reactions carefully, one will notice that this series of reactions results in only a small gain in energy with only eight net ATPs being formed. But an important point is that the reactions *do not require oxygen*. They can proceed even in the absence of oxygen. The reactions that follow glycolysis, however, differ widely based on the presence or absence of oxygen.

TCA, Krebs or Citric Acid Cycle

The TCA/Krebs cycle breaks down pyruvate to carbon dioxide and water and traps the energy in the form of ATPs. It is a complex series of reactions with several key *enzymes and coenzymes* (Figure 7.3). Most of the energy is trapped by the action of *dehydrogenase enzymes* that assist in removing a molecule of hydrogen and adding it to an atom of oxygen to give water (basically a process of oxidation). But it is not a simple process of removing molecular hydrogen and adding it to atomic oxygen. This happens with a chain of events involving the transfer of electrons in the mitochondria through what is called an *"electron transport system."* The net gain from glycolysis and the TCA cycle is 38 ATPs. Several nucleotides assist in this energy transformation: NAD—Nicotinimide adenine dinucleotide (old name DPN, Diphospho pyridine nucleotide); NADP—Nicotinimide adenine dinucleotide phosphate (old name TPN, Triphospho pyridine nucleotide); FAD—Flavin adenine dinucleotide; GDP—Guanosine diphosphate; GTP—Guanosine tri-phosphate.

The oxidation process (removal of H) in the electron transport system goes from NAD to FAD, cytochrome and finally to oxygen for the formation of water. The net number of ATPs synthesized varies depending where the process enters the electron transport system. From the NAD step 3 ATPs are accumulated (6 per glucose molecule); from the FAD step, only 2 ATPs result (4 per glucose molecule). The ATPs accumulated through glycolysis and the TCA cycle are detailed in Table 7.1. In addition to providing the energy, the TCA cycle also provides the carbon skeleton in the form of citric, oxalic, fumaric and succinic acids, all of which occur in considerable amounts in fruits.

The usual substrates for respiration in plant tissues are the carbohydrates and organic acids, which apart from being relatively abundant, generally are utilized in preference to other possible energy sources such as fats, proteins, etc. However, the breakdown products of these other macromolecules also can enter the Krebs cycle upon breakdown by various processes into simple compounds. For example, starch may be degraded by enzymes to glucose and then goes through the usual glycolysis and Krebs cycle. Pectic substances get broken down to galacturonic acid and then to glucose. Fats are converted to glycerol, and

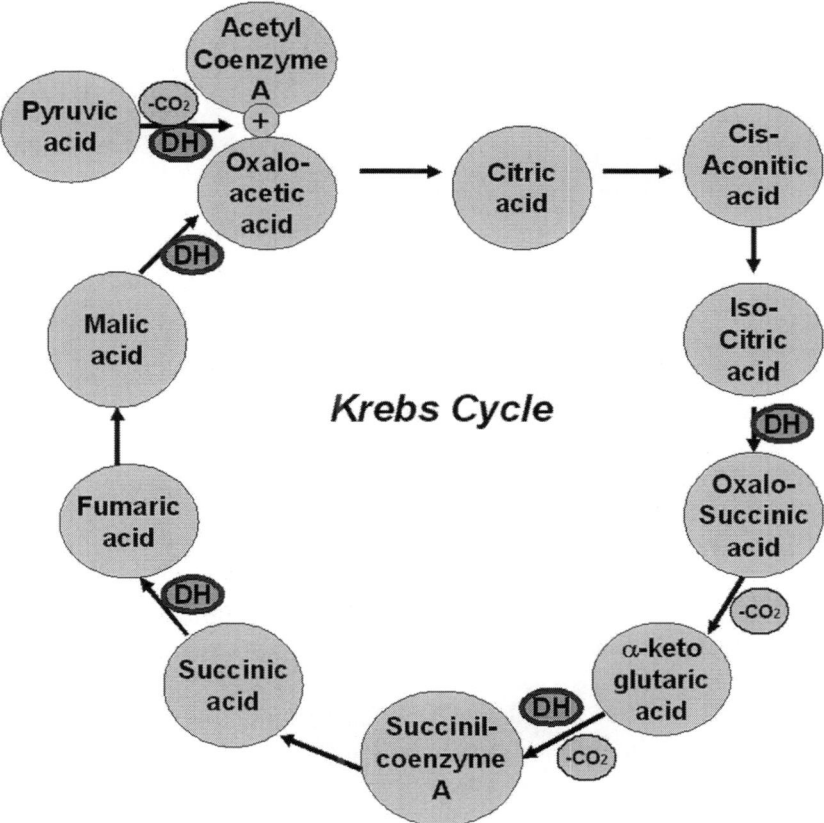

FIGURE 7.3. Krebs cycle.

TABLE 7.1. Summary of Energy Involving Reactions in Aerobic Breakdown of Glucose (Glycolysis and TCA Cycle).

Reaction	ATP
Glycolysis	
1. Glucose to fructose-1-diphosphate ATP converted to ADP plus iP.	
Steps 2 onward are for each 3-carbon fragment. Since glucose can contribute two such fragments, the number of ATP formed for each reaction must be doubled to get what might be expected from glucose. ATPs shown are for one molecule of glucose.	2(-)
2. 1,3-diphosphoglyceric acid to pyruvate: ADP to ATP	4
3. 3-glyceraldehyde phosphate to 1,3-glyceraldehyde DP NAD to NADH + H⁺	6
Sub total glycolysis	8
TCA Cycle	
4. Pyruvic acid to *acetyl CoA:* NAD to NADH + H⁺	6
5. Isocytric acid to Oxalo-succinic acid NADP to NADPH + H⁺	6
6. a-ketoglutaric acid to *succinyl CoA:* NAD to NADH + H⁺	6
7. Succinyl CoA to succinic acid GDP to GTP which can convert ADP to ATP	2
8. Succinic acid to fumaric acid: FAD to FADH$_2$	4
9. Malic acid to oxaloacetic acid: NAD to NADH + H⁺	3
Sub total TCA cycle	30
Net total ATPs formed	38

Note: Conversions through Electron Transport System.
1. Each conversion of NAD to NADH + H⁺ releases 2H⁺ plus 1/2 O$_2$ to form H$_2$O and give 3 ATPs.
2. Each conversion of NADP to NADPH + H⁺ releases 2H⁺ plus 1/2 O$_2$ to form H$_2$O and give 3 ATPs.
3. Each conversion of FAD to FADH$_2$ releases 2H⁺ plus 1/2 O$_2$ to form H$_2$O and give 2 ATPs.

then to pyruvic acid before entering the Krebs cycle. The fatty acids from the fat will enter the citric acid cycle through the acetyl coenzyme A. Proteins are degraded to amino acids and enter the chain at different points like alanine, serine and cysteine via pyruvic acid, aspartic acid, tyrosine, phenyl alanine via oxaloacetate and citric acid.

Pentose Phosphate Pathway

The metabolic scheme outlined above, though probably the most common pathway for carbohydrate metabolism in plant tissue, is not the only one which is known to occur. In some cases the citric acid or TCA cycle may be short-circuited by formation of glyoxylic acid from isocitric acid which in the presence of *acetyl coenzyme A* forms malic acid directly (glyoxalate shunt). The *pentose phosphate pathway* (Figure 7.4) is another route for the metabolism of glucose. The glucose (or hexose) entering the cycle exits as a pentose; in other words, each time

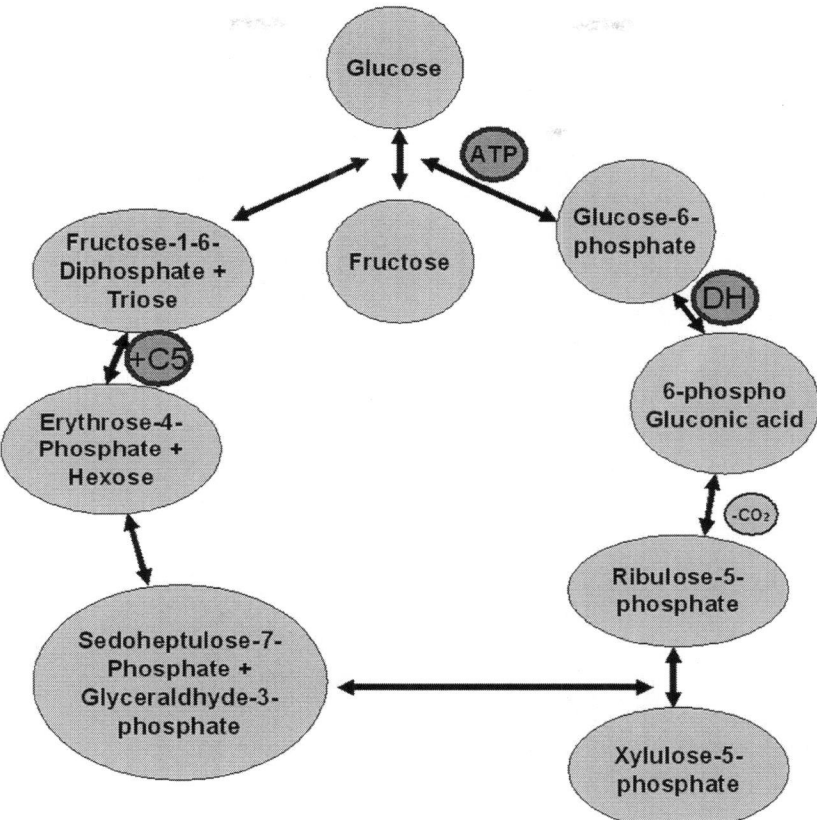

FIGURE 7.4. *The pentose phosphate pathway.*

the cycle is repeated, only one of the six carbon atoms is deleted as CO_2 or only one out of six glucose molecules becomes completely converted to CO_2 and H_2O. It is a physiologically important pathway because it provides 5—carbon skeletons (pentoses) for nucleotide synthesis. It is slightly less efficient than TCA because it only yields 30 ATPs net per glucose molecule $[(6 - 1) \times 6]$ [Figure 7.4].

ANAEROBIC RESPIRATION

As indicated earlier, the first step in the degradation of glucose, which is glycolysis, takes place with or without oxygen and results in the formation of pyruvic acid. In the absence of oxygen, the pyruvic acid is decarboxylated to acetaldehyde, which is then hydrogenated to

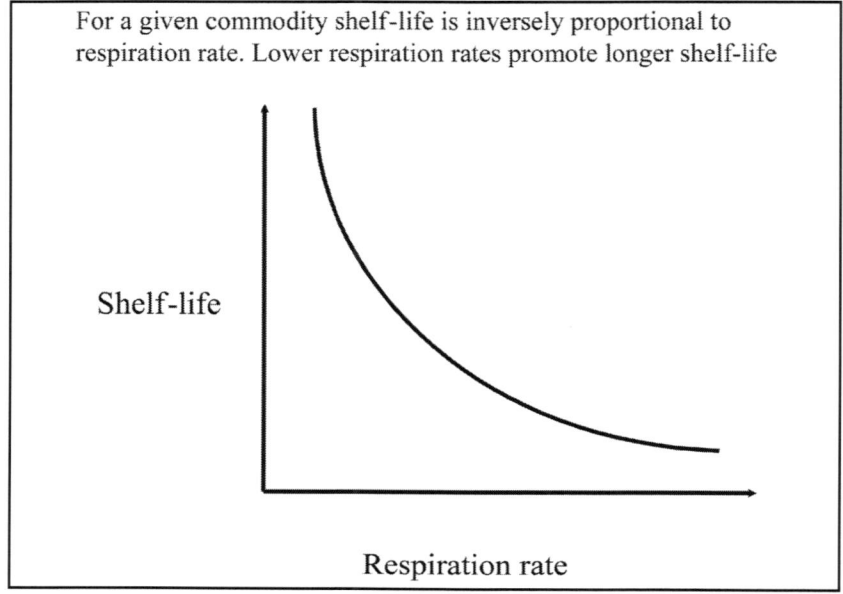

FIGURE 7.5. *Relationship between shelf life vs respiration rate.*

ethyl alcohol, or in the presence of lactic decarboxylase, pyruvic acid is converted to lactic acid. Accumulation of acetaldehyde, ethyl alcohol or lactic acid results in the formation of undesirable flavors.

Pyruvate → *decarboxylase* → Acetaldehyde + CO_2
Acetaldehyde → alcohol dehydrogenase → Ethyl alcohol
Pyruvate → *lactic decarboxylase* → Lactic acid

FACTORS INFLUENCING RESPIRATION

Respiration is the most important physiological activity and has a direct bearing on produce quality. The deterioration of produce quality increases rapidly with the respiration rate (Figure 7.5).

However, respiration rate is not an absolute index of quality deterioration rate because commodities with the same respiration rates can have a different storage life (Table 7.2).

Other factors influence produce respiration: temperature, commodity, variety, maturity, climacteric behavior, availability of oxygen, presence of carbon dioxide, ethylene, growth regulators and others (stresses, injuries, etc.). Some of these are discussed below.

TABLE 7.2. Respiration Rate and Storage Life of Selected Commodities.

Commodity	Respiration Rate (5°C) mg CO_2/kg-h	Storage Life (weeks)
Beans	20	1–2
Brussels sprouts	24	3–4
Cabbage	20	12–16
Spinach	21	1–2

Temperature

Temperature is perhaps the most important factor influencing respiration rate. The respiratory process involves coordinated activity of several enzymes and hence the temperature influence on respiration somewhat resembles that on enzymes. However, since several enzymes act simultaneously, conditions limiting any one enzyme could limit the respiratory activity. Generally, with every 10°C rise in temperature, the respiration rate increases about twofold.

Temperature Quotient of Respiration (Q_{10}): It is an indicator of the temperature sensitivity of respiration rate based on vant Hoff's Law:

Simple Relationship:

$$Q_{10} = R_{(T+10°C)}/R_{(T)}$$

where

$R_{(T)}$: Respiration rate at $T°C$

$R_{(T+10°C)}$: Respiration rate at $T + 10°C$

More General Relationship:

$$Q_{10} = [R_2/R_1]^{\{10/(T_2-T_1)\}}$$

where

R_2: Respiration rate at T_2

R_1: Respiration rate at T_1

While chemical reactions may be characterized by a single Q_{10} over a broad range of temperatures, the Q_{10} concept for respiration and postharvest quality changes is applicable only in narrow temperature ranges because of the complex reactions involved.

Typical respiratory behavior of a fruit during ripening as influenced by temperature is shown in Figure 7.6 indicating a higher rate of respiration as well as hastening of ripening at higher temperatures. Figure 7.6 also illustrates a typical peak respiration rate vs temperature relationship for a ripening fruit. These normally are called climacteric fruits (detailed later), which demonstrates a peak in the respiratory behavior coinciding with the onset of senescence. During this phase, often the ethylene production rate and other metabolic activities also will demonstrate a peak.

Q_{10} for the respiration rate can be calculated either by taking the ratio of respiration rate between two successive temperatures differing by 10°C or by taking the slope of the log (Respiration rate, R) vs Temperature (T) curve (Figure 7.7 is a linear plot of respiration rate vs temperature and Figure 7.8 is a semi-logarithmic plot of Log R vs T). Numerically, Q_{10} can be obtained by dividing the slope of log R vs T curve by 10.

Commodity, Variety and Stage of Maturity

Respiration rates differ for different commodities (Figures 7.9 and 7.10). The rates relate in general to the structure of produce. Leafy and

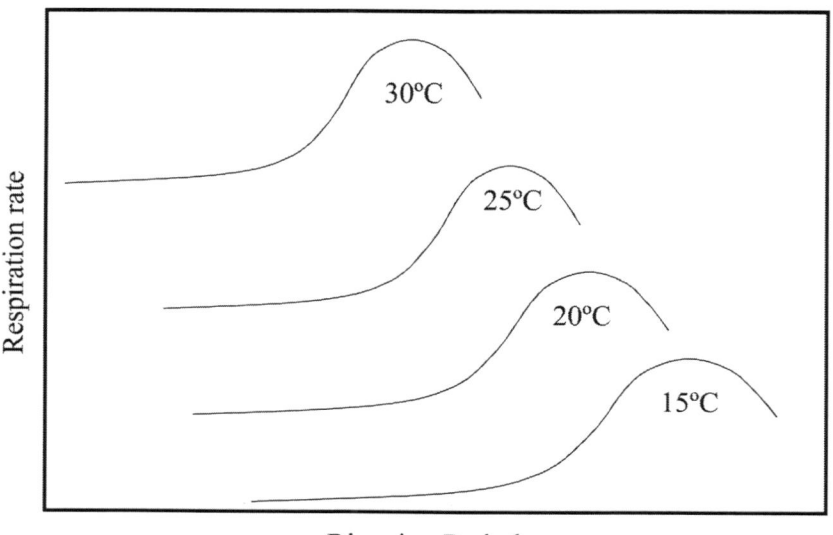

FIGURE 7.6. *Typical respiratory behavior of a fruit as influenced by temperature.*

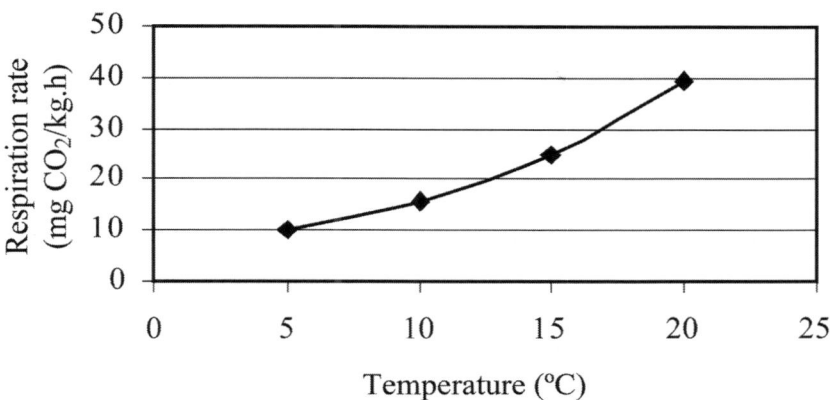

FIGURE 7.7. Respiration rate as a function of temperature.

tender vegetables (spinach, peas, corn, broccoli) have very high rates of respiration while sturdy vegetables like potatoes and onions will have a lower rate. Again, with fruits, the ones with well developed skin (apple, orange, melons, etc.) will have rates of respiration lower than those that are soft skinned (strawberries and raspberries). The respiratory activity is higher at higher temperature for each commodity. In general, with the type of tissue or organ, the rate of respiration is highest with leaves, followed by "regular" fruits and vegetables and is lowest with root vegetables. Generally, smaller size produce has a higher respiration

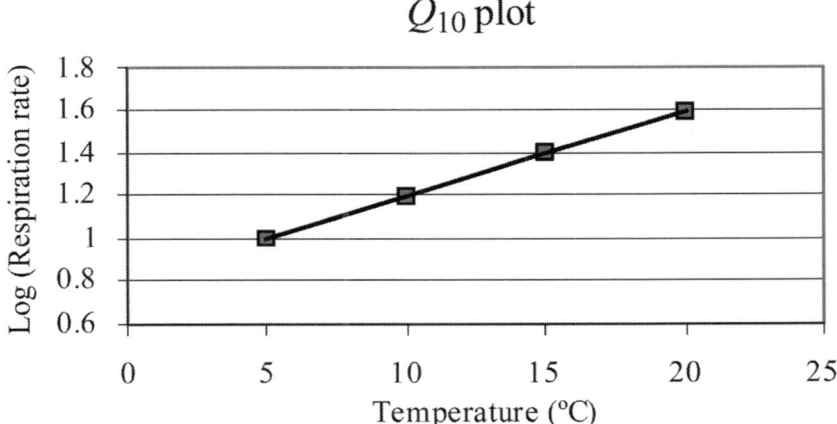

FIGURE 7.8. Respiration rate vs temperature for Q_{10} computation.

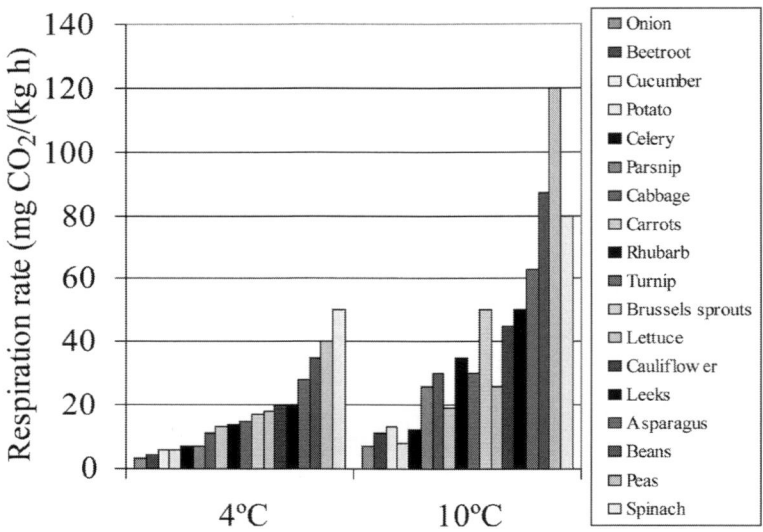

FIGURE 7.9. Respiration rate of selected vegetables at two temperatures.

rate than the larger type. The respiratory activity generally also is high in the developing stage and gradually decreases as the tissue matures (Figure 7.11). In some fruits, the respiratory pattern is characterized by the appearance of a peak with reference to CO_2 or ethylene production

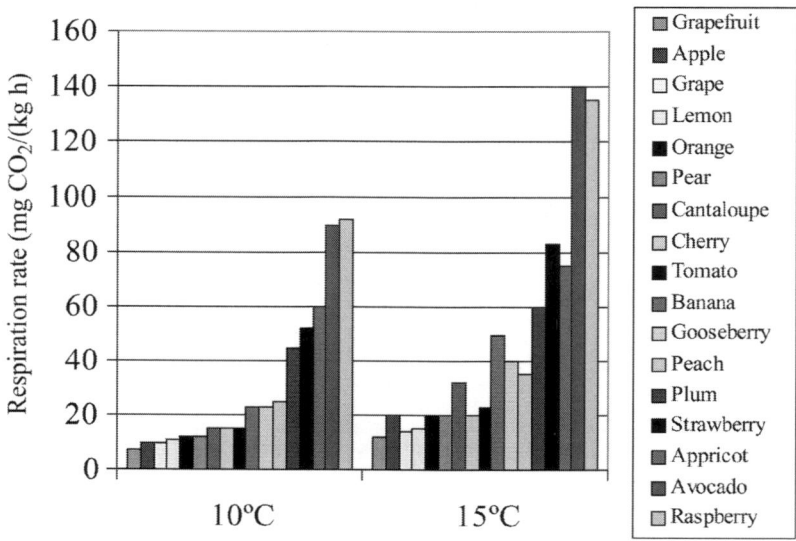

FIGURE 7.10. Respiration rate of selected fruits at two temperatures.

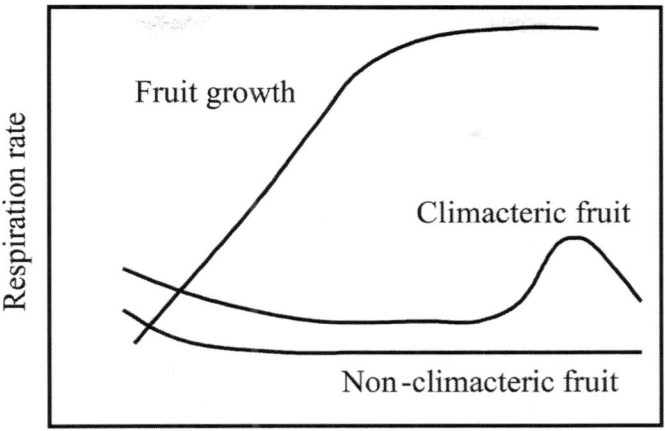

FIGURE 7.11. *Typical fruit growth and respiratory pattern.*

generally coincident with the onset of ripening or senescence (on or off tree) (Figure 7.12). These fruits are called *climacteric fruits*. Others which do not show this characteristic behavior are called *non-climacteric* fruits. Examples of fruits demonstrating climacteric behavior are apples, apricots, bananas, cherimoyas, mangos, papayas, pears, plums, etc. Cherries, grapes, oranges, raspberries and strawberries are exam-

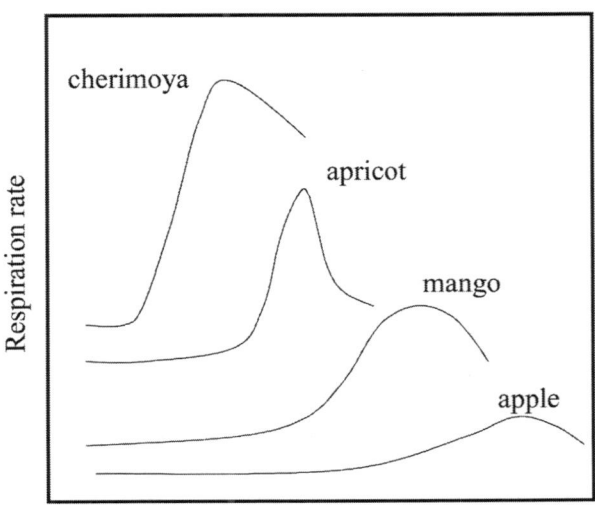

FIGURE 7.12. *Typical respiratory pattern in climacteric fruits.*

ples of non-climacteric fruits. The incidence of the peak respiration rate of climacteric fruits varies with the commodity.

Nature of the Substrate

The nature of the respiration process—i.e., the amount of O_2 consumed and CO_2 released—depends on the type of substrate, as can be seen from the following equations:

Glucose $C_6H_{12}O_6 + 6O_2 = 6CO_2 + 6H_2O$ + Heat (Example 1)

Malic acid $C_4H_6O_5 + 3O_2 = 4CO_2 + 3H_2O$ + Heat (Example 2)

Stearic acid $C_{18}H_{36}O_2 + 26O_2 = 18CO_2 + 18H_2O$ + Heat (Example 3)

While there is a one-to-one correlation between O_2 consumed and CO_2 liberated with glucose as the substrate, the oxygenated substrates such as organic acids (malic acid) will need less oxygen because of the higher oxygen to carbon ratio. Oxygen-limited long-chain fatty acids (stearic acid) will need more oxygen because of their lower oxygen to carbon ratio. The ratio of CO_2 produced to O_2 consumed is termed the "Respiratory Quotient (RQ)" and often is used to identify the nature of the respiring substrate:

Respiratory Quotient (RQ) = CO_2 produced/O_2 consumed

RQ generally is 1 for simple carbohydrates, > 1 for organic acids (1.33 in example 2 for malic acid) and < 1 for fatty acids (0.7 for stearic acid in example 3). Oxygenated substances like organic acids generally have an oxygen to carbon ratio more than 1 in their molecule, and hence require less O_2 for respiration. On the other hand, the long-chain fatty acids have less oxygen per carbon (ratio less than 1) in the molecule, and therefore require more oxygen for the respiratory process. RQ may have some implications in postharvest storage especially in terms of interpretation of respiration rates. Unusually high RQ values with normal substrates may indicate onset of anaerobic respiration, while unusually low RQ values may suggest incomplete oxidation to CO_2.

Availability of Oxygen

Respiration can take place with (aerobic) or without (anaerobic) oxygen. Anaerobic respiration generally is undesirable because of the production of off flavors. Oxygen levels higher than in air (21%) do not necessarily increase respiration rate while levels below 20% decrease the respiration rate. The minimum oxygen level necessary to maintain aerobic respiration in a storage chamber is called the *Extinction Point* (EP) (Figure 7.13).

Storage chambers should have proper ventilation to maintain O_2 levels above the EP. Waxing of produce alters the skin porosity and rates of diffusion of O_2 into and CO_2 out of the produce. In effect, it reduces the availability of O_2 to the tissue thereby permitting micro-respiration, which offers potential for improving produce storage life. Care must be taken to ensure that the diffused O_2 stays above the EP level. Similar precautions must be taken while establishing CA storage facilities. Figure 7.13 indicates that respiration rates decrease up to EP as the O_2 level in the atmosphere decreases, and increase when the O_2 level goes below the EP. The respiratory quotient (RQ) balance is somewhat maintained until O_2 reaches the EP level, but O_2 below EP results in a significant increase in the RQ.

Presence of Carbon Dioxide

CO_2 is a product of respiration, and excess CO_2 favors suppression of respiration (Figure 7.14). Generally, CO_2 concentrations up to 5% have a beneficial effect in minimizing respiration, while in higher con-

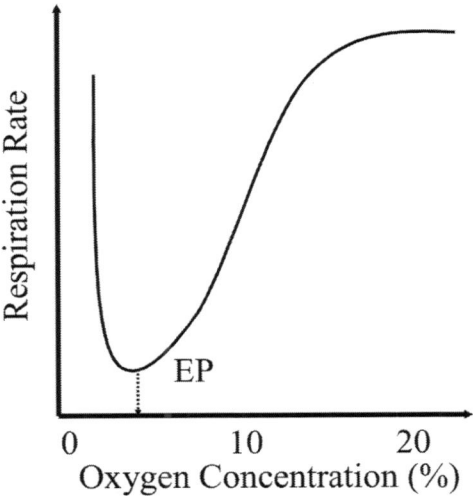

FIGURE 7.13. Graphical representation of respiration rate vs oxygen level.

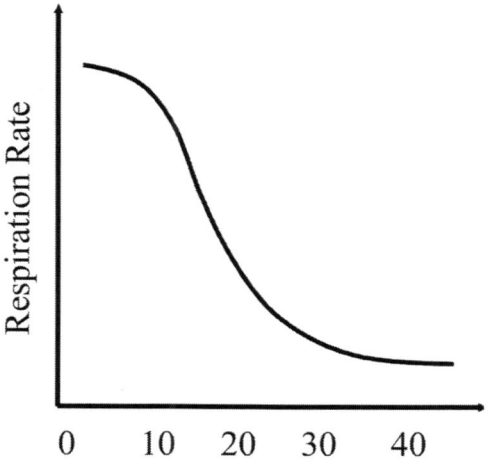

FIGURE 7.14. Graphical representation of respiration rate vs CO_2 level.

centrations CO_2 also has a fungicidal effect. However, the produce must be tolerant of the high CO_2 level. Strawberries can tolerate short-time exposure to high concentrations of CO_2 and hence are generally transported under a high CO_2 environment to suppress mold growth.

Ethylene

Ethylene is a physiologically very active compound even in trace amounts. It has a profound influence on the rate of respiration as well as on ripening of fruits. It is commercially used as a ripening agent and is produced in trace amounts as a result of respiratory activity. Storage systems, therefore, must have appropriate devices (for example, $KMnO_4$, activated charcoal) to scrub ethylene. Typical effects of ethylene on the ripening behavior of climacteric and non-climacteric fruits are shown in Figure 7.15. With climacteric fruits, the respiratory peak (level) usually is not much influenced, but the peak occurs at an earlier date (time shift) with an increase in ethylene concentration. With non-climacteric fruits, however, the rate of respiration is dramatically increased with the application of ethylene.

Growth Regulators

Several growth regulators used in pre- and postharvest applications influence product quality as well as respiration rates. Examples are: delayed or accelerated ripening (Alar); higher yield, greater disease resistance (GA); improved color (Alar); prevention of abscission (NAA); sprout inhibition (MH); and fruit thinning (NAA, Ethephon).

Other Stresses

Injury (chilling injury, freezing injury, physical and mechanical damage) to produce tissue results in hastening of respiration rates. This permits the normally isolated enzymes and substrates present in the food systems to come into contact with each other thereby triggering various biochemical reactions, such as browning, tissue softening, etc.

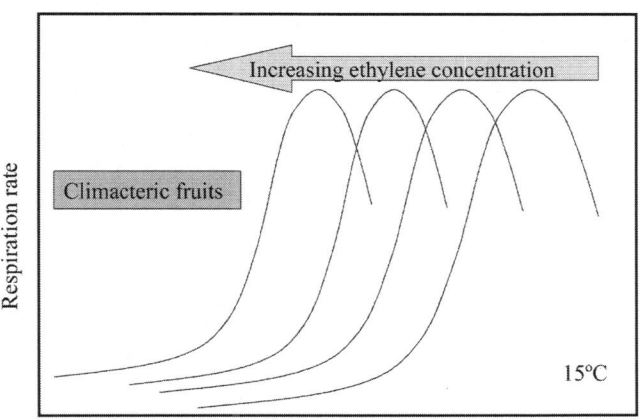

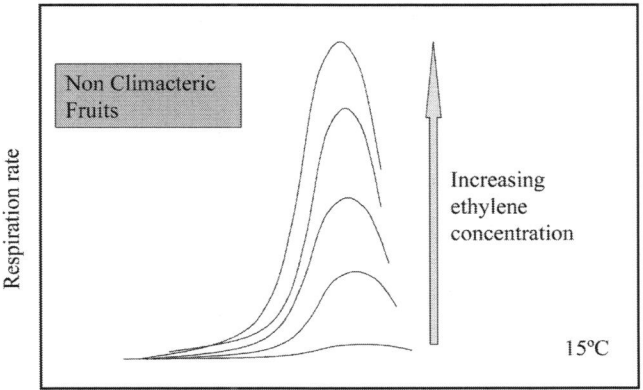

FIGURE 7.15. Typical effect of ethylene on climacteric and non-climacteric fruits.

CONTROL MEASURES TO MINIMIZE RESPIRATORY LOSSES

1. Temperature control
   a. Harvest at cool times;
   b. Cool down and transfer produce to cold store as fast as possible;
   c. Maintain lowest permissible temperature (beware of chilling and freezing injury!);
   d. Maintain the cold storage properly (good air circulation and refrigeration).
2. Maturity
   a. Harvest vegetables at stage appropriate for intended use;
   b. Harvest fruits at mature but at sufficiently preclimacteric stage.
3. Reduce availability of oxygen
   a. Use controlled atmosphere (CA) storage (lowered oxygen level);
   b. Apply appropriate wax where permissible.
4. Add carbon dioxide to environment
   a. Use CA storage;
   b. Use excess CO_2 where permissible.
5. Ethylene
   a. Avoid ethylene-producing chemicals on produce intended for long storage;
   b. Scrub ethylene form storage rooms.
6. Others
   a. Handle the produce with care, to avoid physical damage.

TYPICAL CALCULATIONS INVOLVING RESPIRATION PROBLEMS

Example 1: Respiration rate of potato at 15°C is 12 mg CO_2/kg-h. Calculate the rate of absorption of oxygen, loss of carbohydrate and the evolved heat while storing them at 15°C for one week.

   a. From above conversion factors, we will find that:
   1 mg CO_2/kg-h = 0.727 mg O_2/kg-h
   Oxygen evolved in one week (7 × 24 h) = 12 × 0.727 × 7 × 24
   = 1466 mg O_2/kg

b. From conversion factors, we also will find that:
   1 mg CO_2/kg-h = 61.2 kcal of heat/ton/day
   Heat evolved in one week (7 days) = 12 × 61.2 × 7 = 5141 kcal/ton
c. From the respiration equation, we find that:
   264 g or 264000 mg of CO_2 utilizes 180 g of glucose.
   Therefore, a respiration rate of 1 mg CO_2/kg-h is equivalent to burning up of 180/264000 (= 0.0006818) g of glucose/kg (produce)-h or = 0.0006818 × 1000 × 24 = 16.4 g/ton/day
   In one week, therefore, the amount of sugar lost will be
   16.4 × 7 = 115 g/ton
   In terms of percentages, this would mean a loss of
   115/1000000 × 100 = 0.01%

Example 2: In a study, the respiration rate of carrots at 20°C was found to be 20 mg/kg-h while at 6°C it was 8 mg/kg-h. Calculate Q_{10}.

Data Given:

R_2 = 20 mg/kg-h; T_2 = 20°C
R_1 = 8 mg/kg-h; T_1 = 6°C
Q_{10} ?
Q_{10} = $[R_2/R_1]\{10/(T_2 - T_1)\}$ = [20/8]{10/(20 − 6)} = 1.92

Example 3: In another study, the respiration rate of carrots at 10°C was found to be 10 mg/kg-h and between 5°C and 20°C, it was characterized by a Q_{10} of 2.5. Calculate the respiration rates and heat production rates at 5°C and 20°C.

Data Given:

R_1 = 10 mg/kg-h; T_1 = 10°C
Q_{10} = 2.5
T_2 = 20°C; R_2 = ?
T_3 = 5°C; R_3 = ?
Q_{10} = $[R_2/R_1]\{10/(T_2 - T_1)\}$ or $R_2 = R_1 [Q_{10}]\{(T_2 - T_1)/10\}$ = 10 [2.5]{(20 − 10)/10} = 25 mg/kg-h
Also, $R_3 = R_1 [Q_{10}]\{(T_3 - T_1)/10\}$ = 10 [2.5]{(5-10)/10} = 6.3 mg/kg-h

Example 4: The respiration rate of stored apples at 60°F was 3.5 times that at 35°F. Calculate Q_{10}.

Given:

$R_2/R_1 = 3.5$; T2 = 60°F; T1 = 35°F; Q_{10}?
$Q_{10} = [R_2/R_1]\{18/(T2-T1)\} = [3.5]\{18/(60 - 35)\} = 2.46$

Note that the "10" in the previous equation is replaced by "18" to accommodate the Fahrenheit scale.

Example 5: The high quality storage life of pears at 14°C is 60% less than that at 6°C. Calculate the storage life at 20°C relative to that at 2°C.

Storage life of a given produce at any temperature is inversely related to its respiration rate at that temperature. We can, therefore, rewrite the Q_{10} equation in terms of storage life (S) as:

$Q_{10} = [S_1/S_2]\{10/(T_2 - T_1)\}$
where S_1 and S_2 are storage life at T1 and T2, respectively.

Data Given:

$S_2/S_1 = 0.4$; $T_2 = 14°C$; $T_1 = 6°C$
$S_4/S_3 = ?$; $T_4 = 20°C$; $T_1 = 2°C$
$Q_{10} = [S_1/S_2]\{10/(T_2 - T_1)\} = [2.5]\{10/(14 - 6)\} = 3.14$
$[S_3/S_4] = Q_{10}\{(T_4 - T_3)/10\} = [3.14]\{(20 - 2)/10\} = 7.8$
or $S_4 = S3/7.8$ or 13% of S_3

CHAPTER 8

Post Harvest Physiology of Fruits and Vegetables: Transpiration

INTRODUCTION

IN simple terms, transpiration means loss of water from the produce due to evaporation. Any moist material will lose moisture when exposed to air that is not saturated. Moisture loss from produce occurs through evaporation and vapor diffusion. Transpiration is the second most important factor with respect to quality loss during storage and transportation.

DEGREE OF COOLING ACHIEVED BY TRANSPIRATION

Transpiration is an effective way of keeping a plant tissue cool. The energy needed for the evaporation of water often comes from the produce itself, which brings down its temperature in proportion to the quantity of moisture evaporated. Assuming that all the energy for the evaporation comes from the produce, it is possible to estimate the degree of cooling achieved by transpiration. The following is a simplistic example.

Assume that 1% moisture is lost from 100 kg of an evaporating produce at 20°C. This represents an evaporation of 1 kg of water. Evaporation is a process in which the latent heat of vaporization is absorbed by water resulting in its conversion to vapor or steam. Each kg of water needs to absorb approximately 2260 kJ of heat energy to vaporize. Assuming that this energy comes from the produce, the 100 kg of produce will lose 2260 kJ of energy. Also assuming the specific heat capacity of the produce to be approximately that of water (4.2 kJ/kgC), the temperature difference that the lost energy will bring can be calculated as 2260 kJ/(100 kg × 4.2 kJ/kgC) or 5.4°C. Thus, each one percentage loss in moisture from the surface will bring down the produce temperature by about 5°C.

SIGNIFICANCE OF TRANSPIRATION DURING PLANT GROWTH

1. *The cooling effect:* This was once considered the primary function of transpiration—to keep the plants cool, to dissipate the heat it has absorbed. However, now it is understood, that plants can dissipate heat equally effectively without getting overheated even in the absence of efficient transpiration mechanisms.
2. *Mineral salt and water absorption:* Both these are present together in soil. Both are absorbed by plant roots. Although the driving force for the mineral salts is a more active metabolic response, some certainly enters along with water.
3. *Mineral salt distribution:* Transpiration plays a vital role in the distribution and redistribution of mineral salts to various plant organs.

TRANSPIRATION REPRESENTS ECONOMIC LOSS

Most fresh fruits and vegetables are bundles of water; rather expensive water in fancy containers and packages. Loss of moisture therefore represents an economic loss. This lost water is immediately replenished when the organ is attached to plants. However, when a plant is harvested, no water replacement occurs, which results in a loss of saleable weight.

Moisture loss of 3–4% can make the produce unsellable. Actual moisture loss that a commodity can withstand varies from 3% in leafy tissues to 5% for common vegetables to 10% for onions (Table 8.1).

The moisture loss can seriously influence the product quality with adverse effect on appearance, texture and flavor. Wilting, shriveling, shrinkage, drying, dehydration, desiccation, etc. are results of transpiration activity.

TABLE 8.1. Moisture Loss Upon Which Produce Loses Its Market Value.

Commodity	Range of Critical Moisture Loss (%)	Commodity	Range of Critical Moisture Loss (%)
Asparagus	7–9	Lettuce	3–5
Beans	5–7	Onion	8–12
Beetroot	5–9	Potato	7–9
Brussels sprout	6–9	Peas	4–6
Cabbage	7–10	Pepper	5–9
Carrots	8–10	Spinach	3–5
Cauliflower	6–8	Sweet corn	5–7
Celery	8–10	Tomato	6–8
Cucumber	5–8	Turnip	4–7

TRANSPIRATION OCCURS THROUGH SPECIALIZED TISSUES

The gas and water vapor exchange in fruits and vegetables occurs through specialized surface tissues: the combined action of stomata, guard cells, lenticels, wax, cuticle, cuticular membrane, corky tissue, suberin, etc. The stomata in particular play a primary role in the control of gas and vapor exchange. These are tiny openings in the epidermis with guard cells that open and close continuously. The density of stomates varies from one commodity to another. Leaves may contain as many as 50,000 stomates per cm^2, while an orange peel may contain 1,500 stomates per cm^2.

TRANSPIRATION IS A DIFFUSION PHENOMENON

Transpiration basically is a diffusion process with mass (water vapor) transfer across a surface. It is a dynamic process in which the water vapor from produce diffuses across the produce's surface under the influence of a water vapor pressure gradient. Like all other diffusion processes, transpiration obeys the basic law of diffusion:

Fick's Law:

$$J = AD(\Delta C / \Delta x) / [RT]$$
$$= AD[(Pi - Pa)/x]/[RT]$$

where:

J = gas flux expressed
A = area of the surface
D = diffusivity or diffusion coefficient
C/x = concentration gradient which in this case is the gradient of water vapor pressure $(Pi - Pa)/x$ with x representing the thickness of the surface layer offering resistance to moisture migration
R = Gas constant
T = Temperature (absolute)

Fick's law states that the movement of any gas or vapor in or out of the plant tissue is directly proportional to the partial pressure gradient across the barrier surface, directly proportional to the surface area of the barrier, directly proportional to the diffusion coefficient of the surface, inversely proportional to the temperature and inversely proportional to the thickness of the surface. The main driving force for moisture migration at a given temperature for a given produce is the water vapor pressure gradient, $Pi - Pa$.

PARTIAL PRESSURE, SATURATED WATER VAPOR PRESSURE AND RELATIVE HUMIDITY

Any amount of water vapor present in air exerts a partial vapor pressure that is proportional to its mass; air and water vapor then behave like a mixture of two gases and obey most gas laws related to mixtures of gases.

P_{wi} represents the partial pressure of water vapor in the internal tissue of the produce while P_{wa} represents the partial pressure of water vapor in the air surrounding the tissue. Since most fresh fruits and vegetables contain a large percentage of water, one can expect the partial pressure of water vapor inside the tissue to be approximately the same as the saturated vapor pressure of water at the given temperature as represented by P_{ws}. The saturated vapor pressure of water at a given temperature (P_{ws}) represents the maximum amount of water vapor the air can hold. A further increase in the vapor pressure only results in condensation. P_{ws} can be obtained from tabular data on steam, a shortened form of which is presented in Table 8.2.

$$P_{wi} = P_{ws} \text{ at } T_p \text{ where } T_p \text{ is the product temperature}$$

TABLE 8.2. Saturated Water Vapor Pressure at Various Temperatures.

Temperature (°C)	Vapor Pressure (kPa)	Temperature (°C)	Vapor Pressure (kPa)
3	0.756	40	7.384
6	0.935	45	9.593
9	1.148	50	12.35
12	1.402	55	15.76
15	1.705	60	19.94
18	2.064	65	25.03
21	2.487	70	31.19
24	2.985	75	38.58
27	3.567	80	47.39
30	4.246	85	57.83
33	5.034	90	70.14
36	5.947	95	84.55
		100	101.4

P_{wa} often is related to P_{ws} to get the relative humidity (RH) of air: RH = $(P_{wa}/P_{ws}) \times 100$. The relative humidity has no meaning without reference to the temperature.

RH = $(P_{wa}/P_{ws}) \times 100$ at T_a where T_a is the temperature of air

The difference between the saturated water vapor pressure at a given temperature and the partial pressure of water vapor in the air is called the water vapor pressure deficit (WVPD) of the air.

$$\text{WVPD} = (P_{ws} - P_{wa})$$

When both produce and air are at the same temperature ($T_a = T_p$), $(P_{ws} - P_{wa})$ provides the driving force for moisture transpiration. It is important to take into account the temperature of the produce and that of the air while referring to this discussion. When T_a is not equal to T_p, then P_{wi} (or P_{ws}) should be obtained at T_p; $P_{wi} - P_{wa}$ or $P_{ws} - P_{wa}$ will then provide the driving force for transpiration.

Consider the following examples:

Example 1: Produce at 21°C and air at 21°C and 50% RH (Figure 8.1)

Saturated water vapor pressure at 21°C (from Table 8.1) = P_{ws} = 2.5 kPa

Since RH = 50%, the partial water vapor pressure in air (P_{wa}) = 2.5/2 = 1.25 kPa

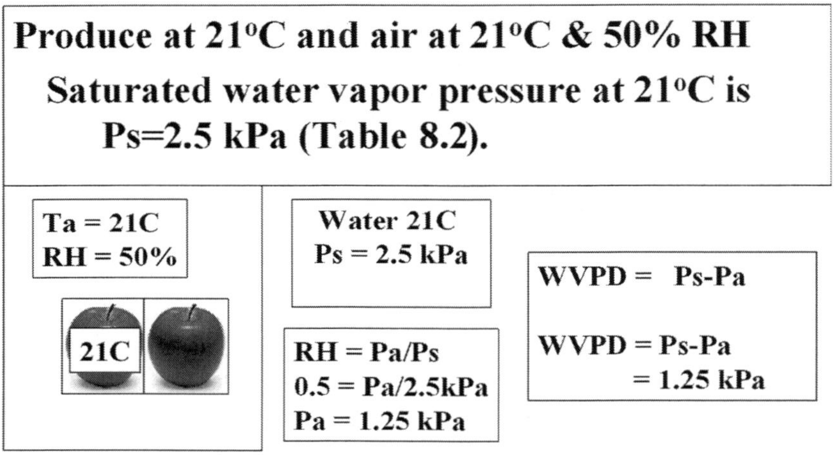

Produce at 21°C and air at 21°C & 50% RH
Saturated water vapor pressure at 21°C is Ps=2.5 kPa (Table 8.2).

Ta = 21C	Water 21C	
RH = 50%	Ps = 2.5 kPa	WVPD = Ps-Pa
21C	RH = Pa/Ps	WVPD = Ps-Pa
	0.5 = Pa/2.5kPa	= 1.25 kPa
	Pa = 1.25 kPa	

FIGURE 8.1. Water vapor pressure deficit calculation.

$\text{WVPD} = P_{ws} - P_{wa} = 2.5 - 1.25 = 1.25$ kPa

The WVPD is positive and therefore the produce will be subjected to transpiration losses.

Example 2: Produce at 30°C and air at 21°C and 50% RH.

Saturated water vapor pressure at 30°C = P_{ws} = 4.25 kPa
Air is at 20°C and 50% RH: therefore P_{wa} = 1.25 kPa
$\text{WVPD} = P_{ws} - P_{wa} = 4.25 - 1.25 = 3.0$ kPa

The WVPD is positive and is 3/1.25 = 2.40 times the WVPD in the previous example. The higher produce temperature will accelerate the transpiration.

Example 3: Consider a warm produce at 30°C entering a cold room at 3°C and 90% relative humidity.

Saturated water vapor pressure at 30°C = P_{wi} = 4.25 kPa
Saturated water vapor pressure at 3°C = P_{ws} = 0.76 kPa
$P_{wa} = P_{ws}$ (air temperature) × RH = 0.76 × 0.9 = 0.68 kPa
$\text{WVPD} = P_{ws}$ (product temperature) $- P_{wa}$ = 4.25 − 0.68 = 3.57 kPa

This WVPD also is positive but is 3.57/1.25 = 2.85 times the WVPD in Example 1.

This means that larger temperature differences between the produce and air will provide a higher WVPD and hence result in larger transpiration losses. Hence, if the produce is not pre-cooled to the storage temperature quickly using an appropriate pre-cooling technique, it is likely to lose a considerable amount of water by transpiration in the storage room during the slow air-cooling process.

Example 4: Room at 3°C and 100% RH (hypothetical, not practical) Produce enters at 30°C

Saturated water vapor pressure at 30°C = P_{ws} = 4.25 kPa

Saturated water vapor pressure at 3°C = P_{ws} = 0.76 kPa

= P_{ws} at 3°C × RH = 0.76 × 1.0 = 0.76 kPa

WVPD = P_{ws} (at 30°C) − P_{wa} = 4.25 − 0.76 = 3.49 kPa

Even this WVPD 3.49/1.25 = 2.8 times the WVPD in Example 1. Although the produce would tend to lose moisture, this moisture cannot be carried by the air because it already is saturated. The resulting moisture simply will stay on the surface causing the produce to "sweat." Such free surface moisture may lead to increased spoilage from molds, etc.

Example 5: Produce at 12°C and air at 21°C and 50% RH.

Saturated water vapor pressure at 12°C, P_{ws} = 1.40

P_{wa} at 21°C and 50% RH = 2.5 × 0.50 = 1.25 kPa

WVPD = $(P_{ws} - P_{wa})$ = 1.4 − 1.25 = 0.15 kPa

This WVPD is 0.15/1.25 which is only 12% of the WVPD at 21°C

Hence, the transpiration is considerably suppressed in this situation.

If the material is pre-cooled further to 10°C, WVPD will be essentially zero and there will not be any moisture loss during short-time exposure to the air at 21°C and 50% RH. However, remember there is a temperature difference between the produce and air, which causes the produce to slowly warm up, and as it does P_{ws} increases causing a positive WVPW, which increases with time promoting moisture loss.

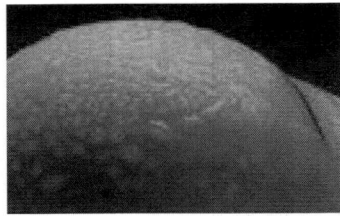
Condensation

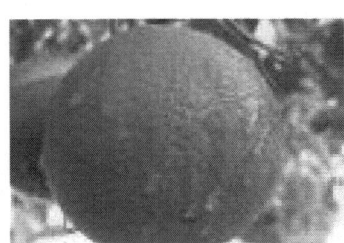

FIGURE 8.2. Condensation examples.

Example 6: Let us look at a different situation when we transfer the produce from the 3°C room to air at 21°C and 50% RH.

The saturated water vapor pressure at 3°C was 0.78 kPa = P_{ws}

The partial pressure of water vapor in air is 1.25 kPa = P_{wa}

WVPD = $P_{ws} - P_{wa}$ = 0.78 – 1.25 = –0.47 kPa which is negative

Here there will be no transpiration; in a strict sense, the produce should gain moisture which might happen to some extent, but in most situations there will be condensation of moisture on the produce surface, similar to that in Example 4. However, in this case, the moisture comes from the air rather than from the produce as was the case in the earlier situation.

Example 7: Curing situation: Curing is a treatment for some roots, tubers and bulbs for extending their postharvest shelf life. This is commonly done for onions, potatoes, garlic, yams, etc. Curing treatment may be applied in the field when conditions are favorable or in a packing house under more controlled conditions (Figure 8.3). The treatment helps the surface layer (periderm) of the produce to dry and to offer resistance to moisture loss and decay caused by bacteria and fungi. Typical treatment conditions are shown in Table 8.3.

TABLE 8.3. Typical Curing Conditions.

Commodity	Temperature (°C)	RH (%)	Time (days)
Potato	15–20	85–90	5–10
Yam	32–40	90–100	4–7
Onion, garlic	35–45	60–75	0.5–1
Cassava	30–40	90–100	4–5

The early part of curing is similar to Example 5 with a cooler produce subjected to warm and dry air treatment. In this case, the objective is to create a relatively drier surface, and the conditions favor this. But this should be done in a controlled way and gradually to prevent excess dehydration. During the second part, when the produce is cooled to storage temperature following curing, care must be taken to prevent moisture condensation (i.e., *sweating*), as the produce gets cooler and the air gets humidified as in Example 4.

Harvesting of roots and tubers can cause damage and skin wounds. They have the ability to heal skin wounds when held at relatively high temperature and humidity for few days. Curing refers to the process of self-healing of wounds, cuts and bruises. There are two steps involved in the curing process: cell suberization which involves the production of suberin and its deposition in cell walls, and formation of corky tissue

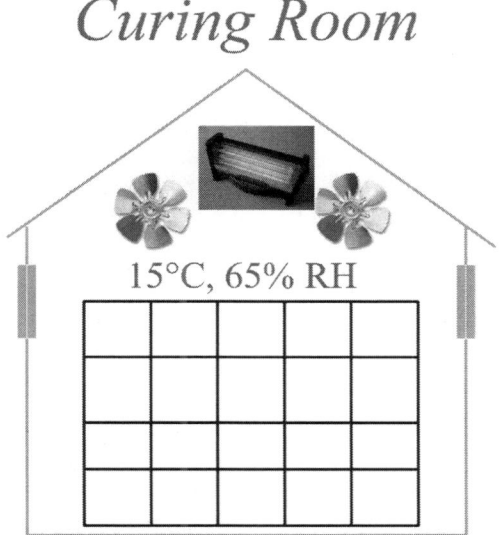

FIGURE 8.3. A typical curing room with a heater, circulating fans and exhaust windows.

in the bruised area. The tissue seals the cut or bruised areas and helps in preventing the entrance of decaying organisms, while also reducing water loss.

Wound healing is important in preventing invasion by pathogens. It also is important in limiting the rate of respiration and water loss. It should be noted that curing involves conditions of high temperature and high humidity within the commodity, which also favor the rapid growth of microorganisms and commodity deterioration. Therefore, it needs to be carried out very carefully and quickly, followed by bringing the produce to proper storage conditions.

Transpiration Coefficient

In the previous examples, we basically looked at water loss from the viewpoint of the water vapor pressure deficit only. In this respect, all produce would behave in a similar fashion and lose moisture in an identical way. But we know that is not true. The surface characteristics of the tissue may be different, sizes may be different, surface area may be different, and these will result in differences in the transpiration rates.

In order to compare the moisture loss from given commodities under comparable storage conditions, another term known as the transpiration coefficient must be defined,

The Transpiration Coefficient (TC) is the quantity of water vapor evaporated at a given temperature by a unit weight of the product in a unit time under a unit gradient of WVPD.

Typical unit of Transpiration Coefficient = mg H_2O/kg produce/MPa WVPD/s

The transpiration coefficient therefore refers to a flux per unit mass of the produce under the influence of a unit WVPD. The produce characteristics, such as area of surface, diffusion coefficient and thickness of the barrier layer, are combined and replaced by the weight of the produce. The weight of the produce can provide estimates of these variables. The transpiration coefficients for fruits and vegetables vary based on their size, shape, surface characteristics, surface/volume ratio, etc. The leafy vegetables have a very high TC due to their large surface area per given weight. The root vegetables also show considerably high TC because they do not possess as efficient a transpiration-preventing structure as that of the aerial organs. During their normal growth, there often is an abundant exchange of moisture between the tissue and soil.

Another convenient way to express TC is by expressing the moisture loss as a percentage of the initial load on a daily basis, especially with reference to storage. Expressed this way, transpiration loss will be a percentage of water lost per day per unit WVPD. Transpiration losses in different commodities vary considerably under different conditions of storage. Expressed as water loss per day per mbar WVPD and at 10°C, it ranges from 0.02% for potatoes, 1.6% for beetroot to 1.9% for carrot roots and 2.4% for bunched carrots. For fruits with skin and peel, it generally is very low, often below 0.05%. With leafy vegetables like lettuce, spinach and watercress, the daily loss can be as high as 10–30%, which means they cannot be left unprotected for even a few hours without serious loss in quality.

FACTORS INFLUENCING TRANSPIRATION

Several factors influence the transpiration process in produce during storage. From Fick's law, it is clear that the surface characteristics (diffusion coefficient and thickness of the surface layer), surface area of the produce, water vapor pressure deficit as well as temperature will influence the mass flux or transpiration of the produce.

Surface Characteristics

Unlike those associated with polymeric films, the diffusion coefficients of fruit and vegetable surfaces are difficult to measure. Hence, this generally is included with the surface characteristics of the material. In this respect, the extent of cutinization of epidermal cells or the extent of suberization of corky tissues determine the transpiration rates. The thickness factor reflects the nature and extent of these protective coatings on the produce surface.

Surface Area

The shape and size of the produce determine the available surface area. Produce with large surface-to-volume ratios will have very high transpiration rates. These also would normally have higher surface-to-volume ratios, and hence when the transpiration rate is expressed on a unit weight basis, the produce with a larger surface/volume ratio will yield higher rates of transpiration.

TABLE 8.4. Examples of Surface/Volume Ratios (cm^2/cm^3).

Leaves	50–100
Cereal grains	10–15
Legumes, small fruits	5–10
Tubers, pome, stone fruits, citrus, banana, onion	0.5–1.5
Cabbage, turnip, yam	0.2–0.5

Water Vapor Pressure Deficit (WVPD), Temperature

WPVD and temperature were discussed earlier with some examples indicating the following general conclusions:

1. At any given temperature (air or produce), the rate of transpiration increases as the relative humidity (or water vapor pressure) of the air decreases, because these conditions favor increase in the WVPD.
2. At any given temperature of air, ahigher produce temperature will increase transpiration rates, again providing a positive WVPD.
3. At any given temperature of the produce and a given relative humidity of air, lower air temperatures (lower than that of the produce) will have a high initial transpiration rate (larger WVPD).

Respiration

Respiration produces heat and moisture. The heat, if not removed efficiently, can increase the produce temperature and hence its transpiration rate. The amount of water produced by respiration is insignificant in relation to the water lost by transpiration.

Air Movement

Air movement plays an important role in storage chambers. On one hand, it assists in better heat transfer to effectively remove the heat of respiration, and any sensible or field heat that is associated with the produce. On the other hand, it also promotes mass transfer in terms of removing the moisture transpired by the produce in its vicinity. This results in further transpiration. If the moisture that comes out of the produce is held in its immediate vicinity, this would help to reduce the WVPD and hence its transpiration rate as would happen, for example, in packaged produce. Both from the heat and mass transfer point of view, higher air velocities increase the rate of heat/mass transfer. In storage systems, a balance is achieved with reference to air velocity between a desirable heat transfer rate and an undesirable mass transfer rate. Moderate speeds of about 50–75 fpm are often employed.

Physical or Mechanical Damage

Wounds, cuts, etc. caused by mechanical or physical damage increase the transpiration rate because they expose the cellular components directly to the environment without the normal protection that previously was offered by the protective tissues.

METHODS OF CONTROLLING TRANSPIRATION LOSSES

1. Maintain as high a relative humidity in the chamber as possible. Several techniques (humidification, jacketed storage) are employed for this purpose.
2. Maintain a low temperature difference between the cooling coils and produce. This will reduce the chances of the cold air being cooled below its dew point. This is achieved by maintaining large cooling surfaces, better refrigeration capacity, moderately high air speeds, proper stacking of produce and prompt precooling of produce prior to storage.
3. Protect from mechanical and physical injuries.
4. Regulate moisture loss by controlling the permeability of the produce tissue. Primarily this is achieved by waxing (apples, citrus, cucumber, bell pepper, banana, etc.). A good wax should have low toxicity, rapid drying characteristics, strong adherence characteristics, high gloss and low cost.
5. Prepackage in polymeric films. Prepackaging is a very effective way of controlling moisture loss by maintaining high humidity levels within the package. Perforations must be provided to allow gas escape. Otherwise CO_2 gas will build up inside. Permeability of film to oxygen is important to keep the required minimum oxygen levels inside the package. The produce is allowed to transpire to accumulate the moisture to the optimal level which will automatically suppress the subsequent transpiration.

PSYCHROMETRY

An understanding of the properties of air water vapor mixtures is required to determine proper postharvest handling and storage conditions. The study of the properties of air and water vapor mixtures is called psychrometry. This section will provide the reader a brief overview of the basics of psychrometry. For additional details, the reader is referred to books on food/chemical engineering (Singh and Heldman, 1993; Charm, 1978; Perry, 1984).

Psychrometric Principles

Psychrometry is the study of the thermodynamic properties of air and water vapor mixtures. The common psychrometric variables are: absolute humidity, relative humidity, dry, wet bulb and dew point temperatures, humid heat, humid volume, etc. A psychrometric chart describes the relationships among these parameters.

A simplified version of the psychrometric chart is shown in Figure 8.4. The mathematical basis for the development of this chart will not be discussed here, but we will look at several examples that describe the principles and illustrate the use of the psychrometric chart in air conditioning applications as required in cold and controlled atmosphere storage environments.

Chart Explanations:

First, let us examine the chart and identify the different psychrometric chart parameters.

1. *Dry bulb temperature:* This is the horizontal axis of the chart. This is the temperature registered by a regular thermometer or any other temperature monitoring device (thermocouple, thermister, RTD, etc.).
2. *Absolute humidity or humidity ratio:* This is the vertical axis of the chart shown on the right hand side. This refers to the moisture content of the air on a dry-weight basis. It is the amount of moisture contained in a given amount of air, expressed on this chart as kg water vapor per kg dry air.
3. *Relative humidity (RH) of air:* RH is represented by the series of curves shown on the chart from 10% to 100%. The line marked

20 represents the 20% relative humidity line while that indicating 80 represents 80% relative humidity. In a way, the chart also gives a measure of the moisture present in the air, with higher relative humidity generally indicating higher moisture in the air. However, since the lines go across the whole horizontal axis represented by dry bulb temperature as well as humidity ratio, RH has no meaning unless it is tied to another psychrometric variable such as dry bulb temperature or humidity ratio. It is always defined with respect to a specific temperature.

Mathematically, relative humidity is expressed as the percentage ratio of the water vapor pressure in the air (P_{wa}) and the saturated vapor pressure (P_{ws}) at the particular temperature in question. Since the vapor pressure of water varies with temperature, a given amount of moisture in the air represents different RH at different temperatures. That is why it always needs to be specified with a temperature.

$$RH = (P_{wa}/P_{ws}) \times 100\%$$

From the above relationship it is clear that the RH maximum can only be 100%, which happens when the water vapor pressure in the air reaches the saturated water vapor pressure at that tem-

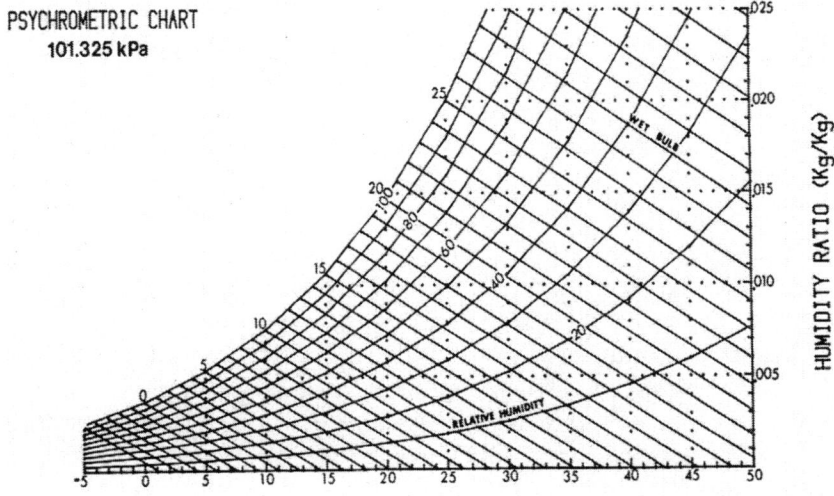

FIGURE 8.4. A typical psychrometric chart in SI units.

perature. This is the condition under which the air can hold the maximum amount of water vapor. From the psychrometric chart, it can be found for any dry bulb temperature by following the vertical line up to 100% RH curve and then determining the absolute humidity or moisture content of the air along the horizontal axis at the absolute humidity or humidity ratio line (Y-axis). It is clear from the chart that air at a higher temperature can hold more moisture than air at a lower temperature. Further, at any given moisture content, RH decreases as temperature increases. If the RH is zero, it means the air is completely dry.

4. *Wet bulb temperature:* Wet bulb is the temperature registered by a thermometer with its bulb surrounded by a wet wick in dynamic equilibrium with the moisture in the surrounding air. These are indicated by the temperature scale on the 100% relative humidity line. The wet bulb lines are represented by the series of slanting parallel lines going down from the wet bulb temperature at 100% RH. At this starting point, the wet and dry bulb temperatures are same, and as the RH decreases, the wet bulb temperature drifts downward away from the dry bulb temperature. Hence, the wet bulb temperature depends on the relative humidity of the air. The lower the RH, the lower is the wet bulb temperature. The following explanation details why and when the wet bulb temperature is different from the dry bulb temperature.

Look at what happens when you have the thermometer bulb surrounded by a moist wick that is constantly kept wet. Allow the air to flow past the wet wick. Assume the air is not saturated with water vapor (RH < 100%). Since the air is not dry, and since the wick is wet and has lots of moisture, the air picks up the moisture from the wick, which results from the evaporation of water. For this evaporation to occur, the water has to pick up the latent heat of vaporization. Since air and the wick are at the same temperature, the heat cannot come from the air. So the heat of vaporization comes from the wick itself. As it gives up this heat, the wick temperature goes down in proportion to the extent of evaporation occurring which again depends on how dry the air is (the lower the relative humidity of the air, the higher the extent of evaporation and therefore higher the lowering of temperature). It can be easily shown that evaporation of 1% water from the wick will lower its temperature by about 6°C. How far down can this temperature go?

Is there a limit? Certainly! As the temperature starts to decrease, a temperature difference is created between the air and the wick. This temperature difference will cause heat transfer from the air to the wick which will cause further evaporation of the moisture or more likely partial replacement of the latent heat coming from the wick required for the evaporation. In the dynamic equilibrium state, the amount of heat flowing into the wick through heat transfer will be equal to the heat required for the evaporation of the moisture. The temperature registered under this condition is the wet bulb temperature of the air.

5. *Dew point temperature:* Dew point is the temperature of air which when cooled will start to condense out its moisture on the cooling surface. As the air is cooled, its partial pressure fraction relative to saturated water pressure increases resulting in an increase in RH (see Figure 8.4). After it reaches the 100% limit, it cannot hold its moisture any longer because the partial pressure will go beyond the saturated vapor pressure and hence the excess moisture will begin to condense out. Classical examples of such situations are appearance of dew in the early morning hours, condensation of moisture on the glasses of cold beverages, condensation of moisture on fruits and vegetables removed from a refrigerator, etc. as illustrated earlier in Figure 8.2. These also are obtained from the temperature scale along the 100% RH line. When air is saturated, the dew point temperatures will be same as the dry bulb and wet bulb temperatures. As the RH decreases, the three temperatures will differ. In order to determine the dew point temperature, follow the horizontal line to the left until reaching the 100% RH line and read the temperature on the line.

6. *State point:* A point on the psychrometric chart representing the psychrometric conditions of the air is the state point. It could be determined by any two psychrometric parameters. Once the state point is fixed, all other properties of the air can be determined from the psychrometric chart.

In Figure 8.5, a state point is shown for the air at 25°C air and 50% RH. This point is obtained as the intersection of the vertical line at a dry bulb temperature of 25°C and the relative humidity curve at 50%. Once this point is identified, the absolute humidity is obtained by moving horizontally right to reach the Y-axis as 0.01 kg moisture/kg dry air. So the moisture content of the air is 1%. Although it is such a small quantity, it plays a dominant role in its moisture exchange behavior in air conditioning applications as well as in cold storages. The wet bulb temperature can be found as 18°C moving along the wet bulb lines to reach 100% RH, and the dew point temperature along the horizontal line to the left is 14°C.

Representation of Different Processes on the Psychrometric Chart

Psychrometric charts can be used to analyze various processes that involve air, such as heating, cooling, humidification, dehumidification or mixing (Figure 8.6). Heating of air by indirect means keeps the original humidity intact (same absolute humidity) and hence moves the state point to the right along the horizontal line, and indirect cooling moves it left. Addition of moisture to the air at a constant temperature represents isothermal humidification moving vertically up and removal likewise is isothermal dehumidification. Adiabatic humidification results in cooling of the air caused by evaporation of water. There will not be any change in heat content of the air since the evaporated moisture stays with the air as vapor. These are represented by moving the state point along the wet bulb line to the left. The adiabatic dehumidification is the opposite, moving the point to the right along the wet bulb line continuously resulting in removal of moisture. Mixing involves air of two

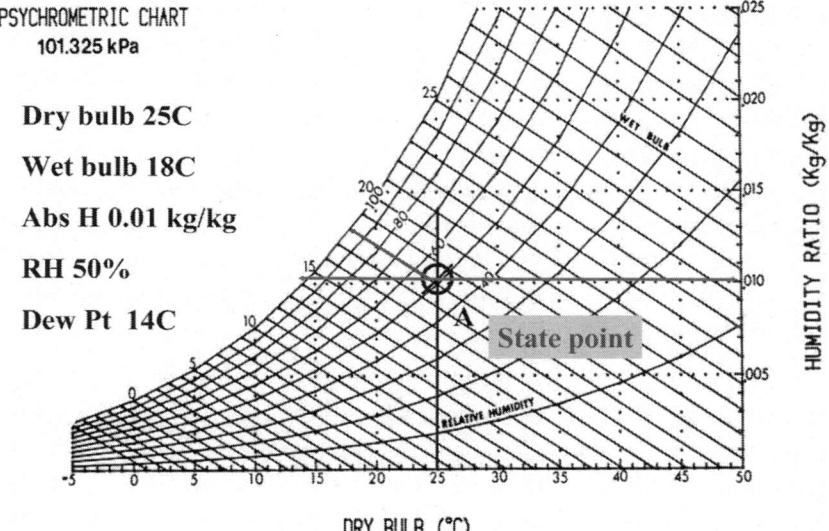

FIGURE 8.5. State point and psychrometric properties.

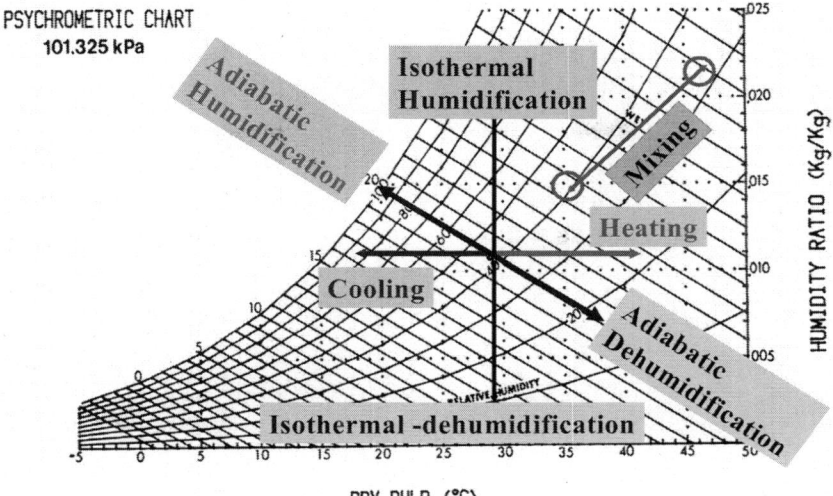

FIGURE 8.6. Representation of various processes on a psychrometric chart.

properties. When mixed, the properties are represented by a state point along the line joined by the state points. The location will be based on the proportion of the two, moving from the center toward the stream of higher proportion.

CHAPTER 9

Cooling of Fruits and Vegetables

INTRODUCTION

What is Cooling or Precooling?

COOLING is a process of removing heat. In other words, it is a process of lowering the temperature of the produce to below the ambient conditions. With reference to produce, therefore, it is a process of removing the *field heat* (heat content of the harvested produce above the storage temperature or the *sensible heat* required to be removed to bring down the produce temperature that existed at harvest to the level maintained at storage).

Precooling technically is no different from cooling. This term was coined almost a century ago to mean cooling of produce prior to shipping. At the turn of the century, shippers found that when loads of warm produce were transported in railway cars to distant markets, it turned ripe or overripe and decayed significantly upon arrival at destination markets in spite of the fact that the cars were refrigerated. This was primarily because the fruits, without the precooling, cooled very slowly in the refrigerated railway car, and spoilage or damage probably occurred even before the cooling process was complete. On the other hand, when these fruits were quickly cooled using some rapid cooling device prior to loading onto rail cars, there was significant improvement in their quality. Soon this became a practice, and shippers were required to precool their produce prior to shipping. Even today precooling still has the same meaning, and includes the cooling procedures prior to storage or processing as well.

Attempts to legalize the definition of precooling, however, failed. In some ways, you could legally claim that your shipment has been precooled by lowering its temperature by a couple of degrees by either storing them under shade or washing them in cold water prior to shipping. Although you could escape legal liability, you would not gain any benefit from such marginal cooling or "precooling." However, a serious shipper is more concerned about the quality of his produce than the legality of precooling regulation, so he would certainly ensure a proper precooling operation even in the absence of legal specifications.

Advantages of Cooling and Precooling

The primary function of precooling is the rapid lowering of the produce temperature. Hence, it will retard all temperature-dependent reactions: respiration, transpiration, microbiological activity, enzymatic activity and chemical activity. Therefore, it results in improved product quality and reduces heat load in storage. As will be clear, precooling is quite demanding on refrigeration, while cold storage is relatively less. Hence, from refrigeration point of view, since precooling takes care of the field heat, the refrigeration design of cold storage needs only to consider the vital heat (heat of respiration) and other heat loads coming from leakages through walls and ceiling, as well as those added during loading and handling of the produce.

Any definition of precooling or cooling should address two important considerations:

1. How much to cool (temperature limits)?
2. How fast this must be done (time limit)?

How Much?

Ideally, the produce must be brought down to the temperature of the environment to which it will be transferred: whether with reference to storage or transportation. In general, the limit is the lowest temperature to which it could be cooled without chilling or freezing injuries. Temperate produce should be preferably precooled to 0–2°C while tropical and subtropical should be cooled to 10–15°C. The starting temperature is the harvest temperature. The lower the field temperature, the better it is from the precooling perspective. Hence, it is advisable to harvest at cool times. Day-time temperatures often are higher than night temperatures; hence, if possible and not prohibitively expensive night harvesting with artificial lighting could offer a significant advantage, and is frequently practiced in California.

How Fast?

Ideally, cooling should be carried out as rapidly as possible, but in practice, as fast as feasible economically and technically. For fruits that can withstand warm temperatures, it could even be achieved slowly over a day or two without a significant loss in quality. However, for perishable fruits like berries, or vegetables like lettuce, celery and spinach, it should be done within a very short time after harvesting and done very rapidly by the best technique available.

Example 1: How much heat must be removed in order to cool one ton (1000 kg) of carrots from 30°C to 5°C?

In order to solve this, we need to know the heat capacity of carrots which tells us how much heat must be removed from one kg of carrots to change its temperature by 1°C.

Heat capacity of most produce generally can be approximated by knowing the moisture content:

$$C_p = 3.35 \, (M_w) + 0.84 \text{ kJ/kg°C}$$

where M_w = moisture fraction: kg moisture/kg produce

Let us assume a moisture content of 90% for carrots, then

$$C_p = 3.35 \, (0.9) + 0.84 = 3.85 \text{ kJ/kg°C}$$

Quantity of heat removed (Q) = Mass × heat capacity × temp difference

$$Q = m \, C_p \times T$$

or

$$Q = 1000 \, (\text{kg}) \times 3.85 \, (\text{kJ/kg°C}) \times (30 - 5)(°C) = 96{,}250 \text{ kJ}$$

This heat removal, which resulted in a change in temperature, is called the *sensible heat* of the produce, because one could sense the cooling effect by feel because of lowering of the temperature. Adding heat will cause the produce to get warmer and removing it will make it cooler, and a person can feel and sense it. Another form of heat is called *latent heat*, which when added to a product does not result in a temperature change. On the other hand, heat transfer results in a change of state, i.e., water changes from solid to liquid (melting of ice) or liquid to vapor (boiling of water) when heat is added or liquid to solid (freezing of water) or vapor to liquid (condensation of steam) when heat is removed. These are referred to as latent heat of fusion (solid-liquid) or latent heat of vaporization (liquid-vapor). This is referred to as latent heat because it results in some latent (hidden) transformation, but one cannot sense its addition or deletion because the temperature remains constant.

In situations where temperature does not change with heat addition or deletion of heat (Q):

$$Q = mL$$

where m is the mass and L is the latent heat.

In order to estimate how fast we need to remove heat or to know how fast we are removing it, we need to estimate the rate of heat removal.

$$\text{Rate } (q) = \text{Quantity } (Q)/\text{time } (t)$$

$$q = Q/t, \text{ J/s} \quad \text{or} \quad W$$

Example 2: In Example 1, if the heat was removed in 2 hours, then what is the rate of heat removal?

The rate of heat removal q can easily be obtained from $q = Q/t$:

$$q = Q/t = 96{,}250 \text{ kJ}/(2 \times 3{,}600\text{s}) = 96{,}250/7{,}200 = 13.4 \text{ kW}$$

In terms of refrigeration requirements, this will mean that the refrigeration unit must be able to remove the heat at a rate of 13.4 kW (operated at 100% efficiency). While designing the appropriate refrigeration equipment, it is customary to at least multiply the capacity required by a factor of 25% for safety.

If we add the safety factor, the refrigeration need will be:

$$13.4 + 25\% = 13.4 + 3.4 \text{ or } 16.8 \text{ kW}$$

Refrigeration needs are normally expressed in tons of refrigeration. There generally is one other consideration while looking at refrigeration systems. In the previous example, if a mechanical refrigeration system is being used to cool the air to remove the heat from carrots, there will be a level of loss (efficiency) in achieving the target. Hence, the refrigeration need will be 16.8 kW at 100% efficiency. Generally, the compressors used in mechanical refrigeration system are rested for a quarter of the time to allow them to cool. This means that the unit needs to produce the desired capacity in 75% of the operating time. Hence, it should be rated at a 75% operating efficiency. As a result, the system requirement will be 16.8/0.75 or 22.4 kW.

Let us further assume there is significant loss in energy transfer (due to heat losses to environment) which accounts for, say, 50% of the original load. These will have to be appropriately accommodated in determining the system configuration.

The calculated original rate of heat removal was 13.2 kW. If there is a heat loss equivalent to 50% of this, the net heat removal rate will be 13.2 + 50% or 1.32 × 1.5 = 19.8 kW. Add the 25% safety, it becomes 19.8 × 1.25 = 24.8 kW. Add the 75% system efficiency and it becomes 24.8/0.75 = 33 kW which will be the system requirement to deliver 13.2 kW of actual use. So the overall efficiency is 13.2/33 × 100 or 40%.

Ton of Refrigeration

A ton of refrigeration is the British unit used to express the nominal cooling capacity of a refrigeration system. It is measured as the heat needed to produce one ton of ice at 0°C in 24 h. Since temperature is constant, the heat being removed to convert water to ice is the latent heat of ice which is 335 kJ/kg (144 BTU/lb). The ton is the short British ton equal to 2,000 lb.

$$1 \text{ ton of refrigeration} = Q/t = (m \times L)/t$$

$$= 2,000(\text{lb}) \times 0.454(\text{kg/lb}) \times 335(\text{kJ/kg}) /(24 \times 3,600\text{s})$$

$$= 2,000 \times 0.454 \times 335/24/3,600 = 3.52 \text{ kW}$$

In the previous example, the refrigeration requirement was 33 kW and in tons of refrigeration it would be 33 kW/(3.52 kW/ton) = 9.4 or approximately 10 tons.

In the earlier examples, we calculated the quantity of heat to be removed by $Q = mCp\Delta T$; and rate by: $q = Q/t$ and refrigeration need by adding a 25% margin to the rate either in terms of kW or tons of refrigeration. In addition, we added some other safety factors like system efficiency and heat loss to environment. However, we did not take into consideration the product-to-product or system-to-system differences. From the above concept, an air cooling system will be no different from a water cooling system or a cryogen-based cooling system. Likewise, there will be no differences in the cooling rate of an apple as compared to a plum or a watermelon, which is not true. Hence, the rate concept is too simple and can be used only to get a global picture of load requirements used in the design of a refrigeration system. These additional considerations will be taken into account when we define the cooling rate and cooling methods in a later section.

Further, we did not take into consideration the type of heat transfer in question. In most cooling and cold storage situations, the nature of heat transfer involved is a combination of convection and conduction. Convection heat transfer is associated with the cooling medium (air, water, etc.), and conduction is within the product. In the next section, we will review some basics related to the different modes of heat transfer.

MODES OF HEAT TRANSFER

Conduction

Conduction refers to the transfer of heat through molecular vibrations or transfer through compacted particles without involving gross movement of particles. This is the common mode of heat transfer in solids. With reference to produce cooling, this will be a typical situation of a tightly packaged container without proper ventilation holes.

Fourier's law of heat conduction states that the rate of heat transfer through a uniform material is directly proportional to the area of heat transfer and the temperature gradient across a unit thickness (i.e., temperature difference per unit thickness) (Figure 9.1):

$$q \propto kA(\Delta T/x)$$

$$\text{or } q = kA(\Delta T/x)$$

$$\text{or } q = \Delta T/(x/kA)$$

$$\text{or } q = \Delta T/R$$

where q is the rate of heat transfer, A is the cross sectional area, ΔT is the temperature difference (driving force), x is the thickness of a slab, k is the thermal conductivity of the material and R is the thermal resistance of the material to heat. The higher the thermal conductivity, the faster it can transfer the heat. When considering heat conduction through walls, each layer including the insulation material offers its own resistance and the total resistance to heat transfer is a combination of all the individual resistances.

Table 9.1 gives examples of thermal conductivities of selected materials. Metals are highly conductive; insulating materials are poor con-

Conduction Heat Transfer

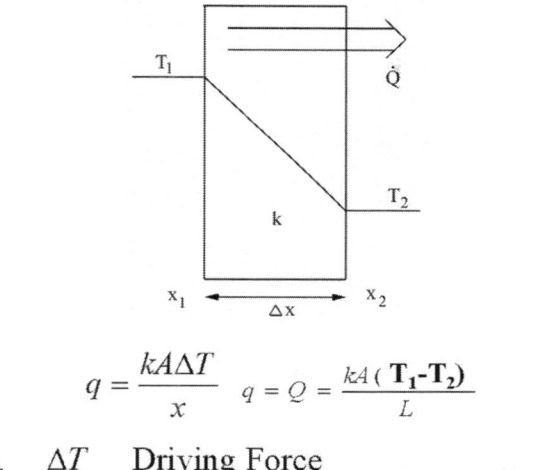

$$q = \frac{kA\Delta T}{x} \quad q = Q = \frac{kA(T_1 - T_2)}{L}$$

$$q = Q = \frac{\Delta T}{R} = \frac{\text{Driving Force}}{\text{Resistance}} \quad \mathbf{R = L/(kA)}$$

FIGURE 9.1. Heat Transfer by Conduction.

ductors. Foods generally have low thermal conductivity. Ice conducts heat about four times faster than water.

Convection Heat Transfer

Convection heat transfer is a result of the gross movement of molecules or particles either by density gradients set up by temperatures or by forced movement of the particles by external means, such as a fan or a pump. Room heating by radiators and the stove top heating of water in a container are common examples of natural convection heat-

TABLE 9.1. Thermal Conductivity of Selected Materials.

Material	k (W/mC)	Material	k (W/mC)
Aluminum	203	Apple	0.39
Brass	98	Beef	0.48
Steel	15	Potato	0.55
Asbestos	0.17	Salmon	0.50
Cork	0.04	Strawberry	0.68
Building brick	0.69	Water	0.54
Wood	0.24	Ice	2.22

ing. Forced circulation of air using fans, as in the case of cold rooms, or pumping of a fluid through a heat exchanger are examples of forced convection, which is used to facilitate faster heat exchange.

Newton's Law of Heating or Cooling states that the rate of convective heat transfer is proportional to the area of heat transfer surface and the temperature difference between the surface and the fluid (Figure 9.2).

Newton's law:

$$q \propto \Delta T$$

$$\text{or} \quad q = hA\Delta T$$

$$\text{or} \quad q = \Delta T/(1/hA)$$

$$\text{or} \quad q = \Delta T/R$$

where q is the rate of convective heat transfer, A is the cross-sectional area of the heat transfer surface, ΔT is the temperature difference (driving force) between the surface and the fluid, k is the surface heat transfer coefficient of the fluid and R is the thermal resistance of the surface to

Conduction through multiple walls

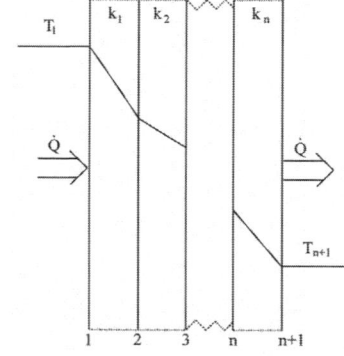

$$R = R_1 + R_2 + R_3$$

$$q = \frac{T_1 - T_{n+1}}{\sum_{i=1}^{n} \frac{\Delta x_i}{k_i A}} \qquad q = \Delta T_1/R_1 = \Delta T_2/R_2 = \ldots$$

FIGURE 9.2. Heat conduction through multiple walls.

Modes of Heat Transfer

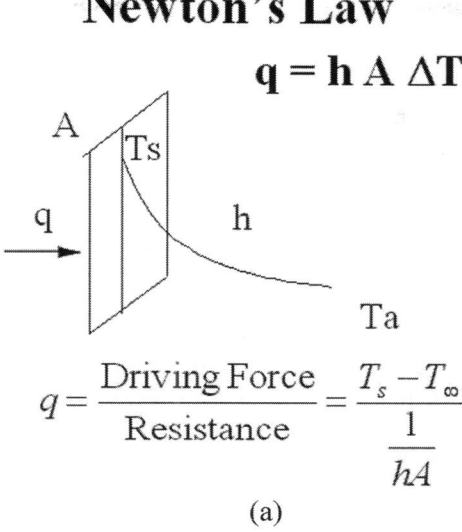

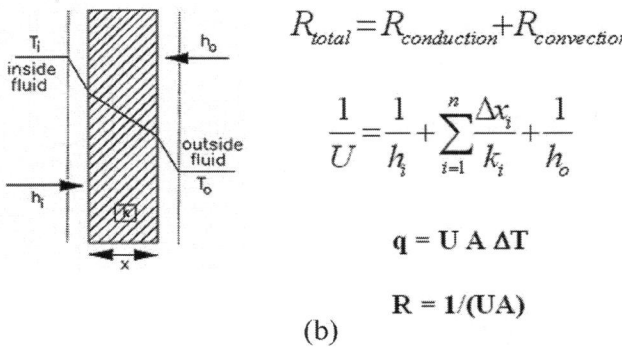

FIGURE 9.3. *Heat transfer by convection (a) and combined conduction-convection (b).*

heat transfer. The higher the surface heat transfer coefficient, the faster it can transfer the heat. Table 9.2 gives examples of surface heat transfer coefficients for selected fluids.

In situations of combined convection and conduction (which mostly is the case in all cooling situations), the convective and conductive resistances become additive:

$$R_{\text{total}} = R_{\text{conduction}} + R_{\text{convection}}$$

TABLE 9.2. Typical Values of Heat Transfer Coefficients.

Fluid	h (W/m^{2}C)
Air (natural convection)	5
Air (forced convection)	20
Water (slowly circulating)	50
Water (rapidly circulating)	100
Liquid nitrogen	400
Boiling water	1,000
Steam	10,000

The conductive resistance generally is considered to be internal resistance and the convective resistance is considered surface resistance. Biot number (Bi) is the ratio of the internal resistance to surface resistance:

$$\text{Bi} = \text{Internal resistance/Surface resistance}$$
$$= \text{Conductive resistance/Convective resistance}$$
$$= [x/kA]/[1/hA]$$
$$= hx/k$$

When Bi < 0.1, the internal resistance is considered to be negligible in comparison with the surface resistance. Therefore, all the resistance occurs at the surface and the material somewhat heats uniformly, i.e., the internal temperature of the product will be the same as the surface temperature. On the other hand, when Bi > 100, the surface resistance is negligible in comparison with the internal resistance, and in this instance, the surface temperature will be the same as the cooling or heating medium temperature for most of the cooling or heating process. Thus, the magnitude of the Biot number has some significance. Biot number also is dimensionless because when the units of h (W/m^{2}C) are multiplied by the unit of x(m), it will give the units for k (W/mC).

Radiation heat transfer does not involve any medium. It travels like any other electromagnetic radiation like light, uninhibited in space. It can be reflected, deflected or absorbed or transmitted through a body. It does not get released until it collides with an object. Transfer of solar energy or infrared heating are examples of radiation heat transfer.

Evaporation refers to transfer of heat due to the exchange of moisture involving the change of state. It involves the transfer of heat to the produce by means of convection, conduction or radiation.

A heat transfer process may involve either a *steady state or an unsteady* state heat transfer condition. Most common heat transfer processes are *unsteady state,* e.g. heating of water, cooling of produce, etc. In unsteady state heat transfer, the temperature at a given location changes with time. The rate of heat transfer also changes with time. In *steady state* heat transfer, the temperature at any given location does not change with time. Hence, *q* remains constant. Heat exchangers are examples where steady state heat transfer exists. The equations described earlier are mostly for steady state systems.

COOLING RATE

Newton's law under unsteady state heat transfer also can be stated as before. The rate of cooling is directly proportional to the temperature difference and inversely proportional to the heat transfer resistance:

$$dQ/dt = -(Ts - Ta)/R$$

where *dQ/dt* is the rate of heat change. It is written as *dQ/dt* because the rate is not constant with respect to time; *Ts* is the surface temperature of the produce, *Ta* is the ambient or cooling medium temperature and *R* is the resistance. The negative sign indicates that the rate is decreasing with time.

In order to simplify, we make an assumption that the temperature of the produce will be same throughout, meaning the surface temperature will be same as the internal temperature at all the locations. This assumption is valid when the resistance to heat transfer at the surface is relatively large compared to the resistance within the produce (Bi < 0.1).

Thus, $dQ/dt = -(T - Ta)/R$ where *T* is the temperature of the produce.

Earlier we used the relationship: $Q = mCp\Delta T$

In terms of small heat changes, *dQ*, this will be: $dQ = mCp\Delta T$

Differentiating with respect to time: $dQ/dt = mCp\Delta T/dt$

Therefore: $mCp\Delta T/dt = -(T - Ta)/R$

Rearranging: $\Delta T/(T - Ta) = -(1/mCpR)\, dt$

Integrating this with respect to time between the time limits zero (Ti) and $t(T)$ gives

$$\ln(Ti - Ta) - \ln(T - Ta) = -(1/mCpR)(0 - t)$$

$$\text{or} \quad \ln[(Ti - Ta)/(T - Ta)] = (1/mCpR)(t)$$

$$\text{or} \quad [(T - Ta)/(Ti - Ta)] = e^{-(1/mCpR)t}$$

The left hand side basically has only temperature-related terms related to the temperature difference between the produce and the cooling medium existing at any given time t, which is $(T - Ta)$ and the initial difference in temperature, $(Ti - Ta)$. This ratio of temperatures is called the residual temperature ratio. This ratio will initially be 1.0 and drops down logarithmically with reference to time. So, in a strict mathematical sense this ratio can never reach zero, but in most practical situations, since we can hardly differentiate temperatures of less than 0.1 degree, when the ratio gets small, we assume the cooling to be complete.

Newton's Law therefore states that the temperature difference between the produce and the medium varies exponentially with time.

The term within the parentheses, which represents the negative slope of the curve when $\ln(T - Ta)$ is plotted against time is called the cooling rate or cooling coefficient (CR or CC) (Figure 9.4).

The equation for the slope of the above curve can be rewritten as:

$$\ln[(T - Ta)/(Ti - Ta)] = -(CR)t$$

Therefore

CR = negative slope and equal to $1/(mCpR)$

The cooling rate contains the resistance R offered to the transfer of heat from the medium and the properties of the produce such as mass and the heat capacity. The resistances are basically additive. So if we have three or four resistances to heat transfer such as a film package, a retail container, and a master container in a bulk container, the heat will have to be removed through all these. The individual resistances will be dependent on their respective thermal conductivities, thickness and area.

The cooling rate may therefore be taken to represent the overall index of the performance of the system under a given condition. The cooling rate is somewhat more difficult to comprehend. A more simple parameter, which is related to CR(CC), is defined for this purpose. This is called the *half cooling time*.

Semi-logarithmic cooling rate curve

$\ln[(T - Ta)/(Ti - Ta)]$ vs Time

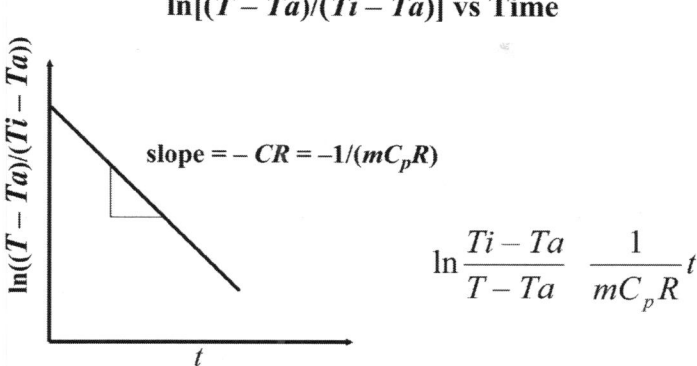

FIGURE 9.4. Cooling rate curve.

Half Cooling Time

Half cooling time is defined as the time interval which will reduce the temperature difference between the product and medium by half [Figure 9.5(a)]. It can be recognized from the figure that the temperature difference vs time curve is a smooth and symmetrical curve demonstrating a semi-logarithmic decline in the temperature difference. Although shown as the time required from 50°C (initial temperature difference) to 25°C (half of the initial), the half cooling time can be computed as the time interval from the temperature difference existing at any time (t) to a time at which the temperature difference reduces to one half of that at time t. From 9.5(b) it can be easily recognized that there can be many half cooling times in a cooling curve, each reducing the temperature difference to one half from the initial. Hence, after the first half cooling time, we accomplish 50% (or 1/2) of the temperature difference; after 2 half cooling times, 75% (or 3/4) will be accomplished, and so on. One also can realize that by knowing the half-cooling time and the initial conditions of cooling (i.e., the produce temperature and the cooling medium temperature), one could calculate how long it will take to cool the product to any given temperature.

Example 3: In a hydro-cooler operating at 2°C, carrots are cooled from 22°C to 10°C. Calculate how long it will take to reach 4.5°C. Assume a half cooling time of 5 min.

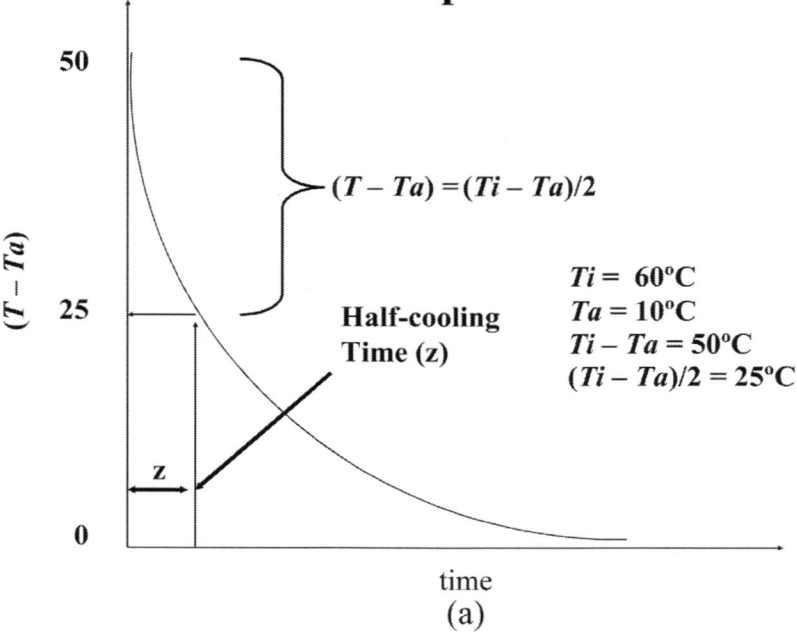

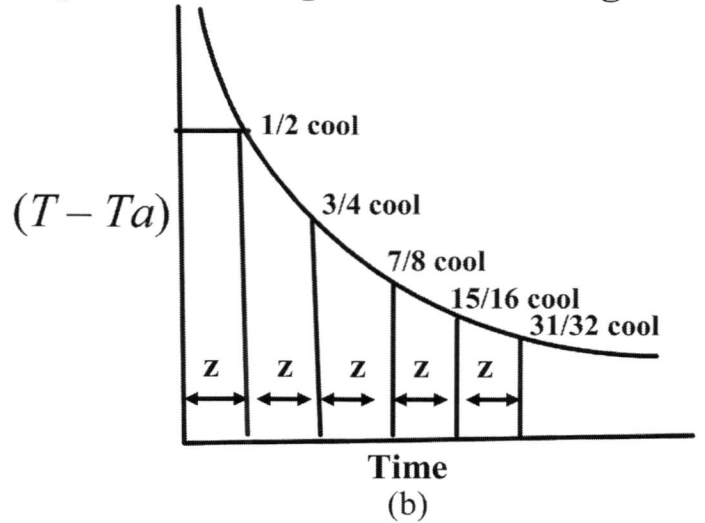

FIGURE 9.5. (a) Half-cooling time definition; (b) Half-cooling time concept: Multiple half-cooling time in a cooling curve.

We know:

- The initial product temperature = 22°C
- Initial temperature difference = 22 – 2 = 20°C
- Target temperature difference = 4.5 – 2 = 2.5°C

After the first half-cooling time, the temperature difference will be 20/2 = 10°C.
After the second half-cooling time the temp difference will drop further to 10/2 = 5°C.
After the third half-cooling time the temp diffifference will drop to 5/2 = 2.5°C

Since we are interested in a target temperature difference of 2.5°C, this stage could be reached after three half-cooling times of 5 min or 3 × 5 = 15 min is the cooling time.

This is a longhand method of solving the problem. If we don't have the target temperature difference matching with the temperature difference at the end of a half-cooling time, the result will have to be interpolated. We could use the equation from Newton's law to get this. We derived the relationship:

$$\ln[(T - Ta)/(Ti - Ta)] = -(CR)t$$

When $t = z$ (which is the half-cooling time), then the temperature difference existing will be $(T - Ta) = (Ti - Ta)/2$.

Therefore

$$\ln[\{(Ti - Ta)/2\}/(Ti - Ta)] = -(CR)z$$

or

$$\ln[1/2] = -CR(z)$$

or

$$z = 0.693/CR$$

In this example, therefore, $CR = 0.693/z = 0.1386$ min

We now know:

$$T = 4.5°C, Ti = 22°C, Ta = 2°C \text{ and } CR = 0.1386$$

Therefore

$$\ln[(4.5 - 2)/(22 - 2)] = -0.1386t$$

$$t = 15 \text{ min}$$

Note the following also:

when $t = z$, $T - Ta = (Ti - Ta)/2$
when $t = 2z$, $T - Ta = (T - Ta)/4$ or $(Ti - Ta)/2^2$
when $t = 3z$, $T - Ta = (T - Ta)/8$ or $(Ti - Ta)/2^3$
when $t = 4z$, $T - Ta = (Ti - Ta)/16$ or $(Ti - Ta)/2^4$

Therefore, when

$$t = nz, \; T - Ta = (Ti - Ta)/2^n$$

This form of the equation can be used to calculate the produce temperature after a given time. In other words after a cooling time of t (which is equal to $n \times z$), the resulting difference in temperature between the produce and the medium will be: $(Ti - Ta)/2^n$, and the produce temperature therefore will be $T = (Ti - Ta)/2^n + Ta$.

Example 4: Using Newton's law, construct a cooling curve for carrots with the following data: $Ti = 28°C$; $Ta = 2°C$ and $z = 7$ min

We can calculate the temperatures at various times using

$$T = (Ti - Ta)/2^n + Ta \quad \text{where} \quad n = t/z$$

In order to find the time when the produce will reach a given temperature, we can write the above equation as:

$$Ti - Ta = (Ti - Ta)/2n$$
$$\text{or} \quad (Ti - Ta)/(Ti - Ta) = 1/2n$$
$$\text{or} \quad \log[(Ti - Ta)/(Ti - Ta)] = n\log(1/2)$$
$$\text{or} \quad n = \log[(Ti - Ta)/(T - Ta)]/\log(2)$$

and since, $t = nz$

$$\text{time} = z\log[(Ti - Ta)/(Ti - Ta)]/\log(2)$$

TABLE 9.3. Calculated Temperatures at Various Times.

Time (min)	n = t/z	Temperature (°C)
0	0	28
5	0.71	17.8
10	1.43	11.7
15	2.14	7.9
20	2.86	45.6
25	3.57	4.2
30	4.29	3.3

TABLE 9.4. Time Taken to Reach Different Temperatures in Example 3.

Temperature (°C)	time = z log[(Ti – Ta)/(T – Ta)]/log(2) (min)
25	1.2
20	3.7
15	7.0
11.7	10
10	11.9
5	21.8

Note: All examples in the table can be solved using the first relationship: $\ln[(T-Ta)/(Ti-Ta)] = -CR(t)$.

Example 5: Use the equation $t = z\log[(Ti - Ta)/(T - Ta)]/\log(2)$ to calculate the time in Example 3:

Data we had:

$$z = 5 \text{ min}, Ti = 22°C, Ta = 2°C, T = 4.5°C \text{ and } t = ?$$
$$Ti - Ta = 22 - 2 = 20°C$$
$$T - Ta = 4.5 - 2 = 2.5°C$$
$$(Ti - Ta)/(T - Ta) = 20/2.5 = 8$$
$$n = \log[(Ti - Ta)/(T - Ta)]/\log(2) = \log(8)/\log(2) = 3$$
$$t = nz = 3 \times 5 = 15 \text{ min}$$

Example 6: Use the data from Example 4 and calculate the times to reach selected target temperatures (see Table 9.4).

We have $Ti = 28°C$; $Ta = 2°C$ and $z = 7$ min
Equation to be used: $t = z\log[(Ti - Ta)/(T - Ta)]/\log(2)$

Factors Affecting the Cooling Rate/Half-cooling Time

Half-cooling time or the cooling rate is a measure of the rate of cooling of a product under a given set of conditions. All product and process related components that affect the rate of heat transfer will affect the rate of cooling. Some examples of these are given below:

Product-related factors affecting the half cooling time:
- Product type, shape, size.
- Product thermo-physical properties which usually are affected by composition of the product.
- Packing and packaging of the product.

System-related factors affect the half-cooling time and include all factors that affect the heat transfer coefficient associated with the medium:

- Type of cooling medium.
- Flow rate of the cooling medium.
- Flow contact and pattern within the cooler and within the package.
- Product load, product stacking.

Product temperature (initial and target) and medium temperature do not affect the half-cooling time, unless they affect the thermo-physical properties of the product or heat transfer coefficient of the cooling medium. They do affect the cooling time, however, in a highly significant way. Lower ambient temperatures will bring down the cooling time; higher product temperatures will require longer cooling time, etc. Table 9.5 gives typical values of half-cooling times under different conditions.

Cooling Procedures

Before discussing the different methods of cooling, we should recognize that cooling and cold-storage are two separate operations based on different principles. Their requirements with respect to refrigeration needs, medium floor pattern and speed of operation are considerably different. Let's use an example to understand the differences in their refrigeration requirements:

TABLE 9.5. Examples of Half-cooling Times for Cooling Procedures.

Commodity	Procedure	Half-cooling Time (min)
Apples	Forced air	50
	Hydro cooled	25
Nectarines	Room cooled	300
	Forced air	50
	Hydro-cooled	15
Cherries	Forced air in:	
	open tray	20
	wood container	40 (without stacking)
	stacking with vents	80
	without vents	150

Example 7: Compare the refrigeration demand needed to cool one ton (1000 kg) of carrots from 30°C to 5°C in 2 hr with the cold storage demand to compensate for the heat of respiration (respiration rate = 5 mg CO_2/kg-h) (heat losses through the building and other refrigeration requirements are not considered in both cases).

In Example 1, we calculated the quantity of heat we must remove to cool one ton (1000 kg) of carrots from 30°C to 5°C as 96,250 kJ and in Example 2, we calculated that if the heat was removed in 2 hours, then the rate of heat removal is 13.4 kW.

Heat of Respiration = Respiration rate (in mg CO_2/kg-h) × 61.2 kcal/ton/day

Heat of Respiration = 5 × 61.2 or 306 kcal/ton/day

= 306 kcal/ton/day × [4.187 kJ/kcal] × [day/24 × 36(

= 0.0148 kJ/ton-s or 0.0148 kW/ton

Since 1 ton = 1000 kg, the heat removal rate from 1000 kg of carrots is 0.0148 kW

Compare this with the heat removal rate (calculated earlier) for cooling of 13.4 kW!

Cooling/Cold storage = 13.4/0.0148 ~ 900

Hence the refrigeration requirement for cooling is almost three orders of magnitude higher than for cold storage. In the cooling process, there is a very high initial demand because of the need to remove a large heat load in a short time. On the other hand, cold storage is intended for keeping the pre-cooled produce for a long term. As will be detailed later, storage's main demand components are: the respiratory load, heat leaks from walls, ceilings, floor, opening of the doors, equipment, people working, fans, and cycling of the external air (venting and intake).

COOLING/PRE-COOLING METHODS

Several methods are available for the general cooling of produce: room cooling, forced air cooling, hydro cooling, vacuum cooling, etc. There also are special techniques used for cooling packaged materials and those used in transit/display cabinets: top icing, package icing, liquid nitrogen cooling, dry-ice cooling, etc. These will be described in this section.

Room Cooling Techniques

Simple Room Cooling

A simple room cooler consists of an insulated room with an appropriate mechanical refrigeration system and fans for air circulation. Because of its simplicity, it is a fairly inexpensive system. Cooling involves placing produce containers in the cold room. Cooling is mainly achieved by moving the cold air around the containers. For efficient operation, containers need to be placed in an orderly fashion with 12–18 inch spaces between them, so that cold air from the fans at the ceiling level flows over the produce containers and moves down to make contact with the containers. Proper stacking of containers is essential for optimal cooling as well as efficient filling of the room. Proper spacing must be provided for air circulation. If field containers are used, the fork lift openings should be clear and free of dirt since these are generally used as air channels. Also, large void spaces in the room should be avoided in order to prevent the air from taking the path of least resistance by bypassing the containers. Pallet frames are used for stacking of containers to provide better load distribution and to allow for air circulation. An air flow rate of 200–300 cubic feet per minute generally is used in the system. Higher flow rates will promote better cooling but, if the same room is used for storage, this may result in excessive desiccation of the produce. If used for cooling and storage, appropriate relative humidity (90–95%) also needs to be maintained. Generally the cooling room and refrigeration equipment are designed in such a way that produce cooling is completed within 24 hours. Since much larger refrigeration equipment is needed to efficiently cool down the produce than to maintain the produce at the cool temperature, these systems are always built with compromises.

Advantages claimed for the simple room cooling include: Produce is cooled and stored at same place, involves less handling, design and operation generally are simple, the system is based on moderate refrigeration load (not too high and not too low), and only moderate air speeds are needed to facilitate the cooling. Disadvantages include: cooling generally is very slow, it is unsuitable for produce which does not tolerate slow cooling, and desiccation can occur if higher air speeds are used because slower speeds prolong cooling.

Several modifications have been suggested to the simple cold rooms to make it more cost effective and cooling efficient:

- *Ceiling jets:* In this modification, open cones are attached to a pressurized false ceiling. The cold air is directed downward from these jets. Floors are marked for placement of pallets in relation to the jets. Cooling achieved is considerably faster because of the forced movement of the cold air.
- *Cooling bay:* A large cooling room is divided into several smaller cooling bays. Each bay is separately and independently controlled for both temperature and air flow rate. Smaller quantity of produce is handled in each of these bays. Since differential air flow rates are possible (higher rates during pre-cooling and lower rates during storage), they can be operated efficiently. Each bay is separate and not affected by other operations. Product handling is minimized. Elaborate control systems and structural modifications can be achieved, but will require more capital.

Forced-air Cooling

Forced-air cooling rooms are specifically designed to promote faster cooling times. These use forced circulation air flow through the containers rather than around them. This also is called pressure or suction cooling. It produces a higher pressure differential which results in a higher air flow rate. The high-velocity cold air is forced through the container. Generally flow is based on a suction principle, which is more efficient than blowing the air onto the containers. Converting simple room coolers to forced-air cooling is practical and feasible. There are many variations of forced-air cooling that fit specific container needs.

The success of forced-air cooling depends on moving the cold air through the container and around its contents. It is therefore important the container have properly designed ventilation ports that allow movement of air through them. It also is important to properly stack the containers and align their vents in order to be directly in the path of the air suction and to prevent either short circuiting of the air flow or blocking. Generally, the operation involves creating a suction tunnel from which the air is sucked out onto the cooling surfaces of the refrigeration system. The suction tunnel is constructed by properly placing the containers with their vents aligned into the tunnel. When the air from this tunnel is drawn out, it will suck the cold air from the room through the container vents, thereby allowing it to move past the produce. This facilitates better and faster air removal from the produce. The top surface of the tunnel usually is sealed with a tarpaulin to allow air in only through container vents (Figure 9.6).

Many variations of pressure cooling system exist. The common features include: air flow through the container, not around, more efficient contact, and faster heat removal for faster cooling. All techniques make use of a fan/blower to pull the air through the produce. The air is then pushed onto the cooling coils. Negative pressure created between the fan and the container by appropriate stacking and blocking of open spaces will result in a suction of cold air from the room through package vents. Important factors to be considered are air flow rate, temperature, package venting and stacking. All types of produce can be cooled using forced air systems.

The containers used for forced air cooling have to be properly designed to allow air circulation through them. Larger and multiple vents will promote better air transfer, but they will weaken the containers as well. It generally is recognized that a 5% vent area allows the cooling time to be reduced by 25% while reducing the package strength by about 2–3%. Vents close to the edges also cause considerable weakness in the package. A properly designed container will have approximately 4–5% vent hole space on the sides, away from the corners, but aligned on the two opposite sides (Figure 9.7).

Figure 9.8 shows a commercial forced air cooling facility, called a tunnel cooler, which is similar to the concept just described. Cold wall and serpentine cooling are some variations of forced air cooling systems.

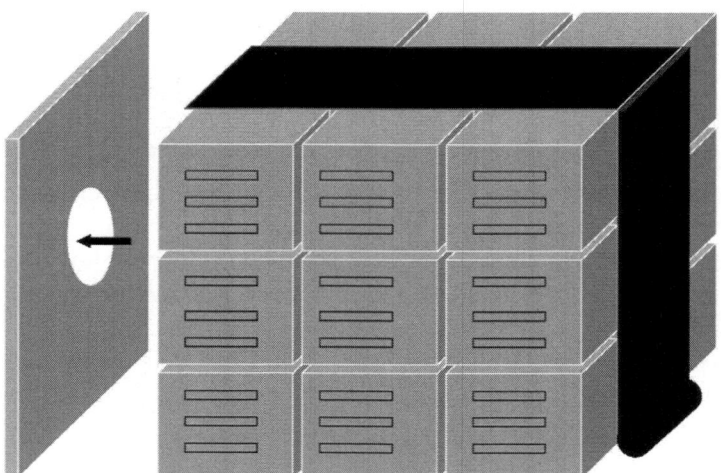

FIGURE 9.6. Schematic representation of forced air cooling.

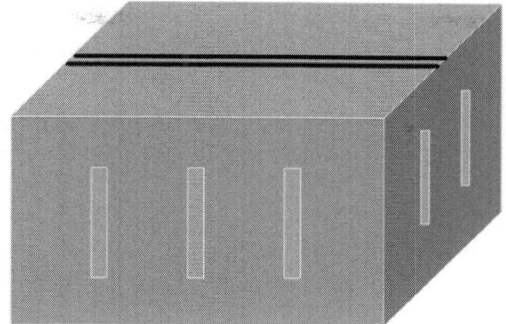

FIGURE 9.7. A typical package with vertical side vent ports.

- *Cold wall:* The cold wall is similar to a cooling bay and consists of an exhaust fan that draws air from the cold room and directs it over the cooling surface. The wall is built with a damper system that only opens when containers with openings are placed in front of it. The fan pulls cold room air through the container and contents, thereby cooling the produce. The exhaust air is forced onto the cooling area of the refrigeration unit prior to letting it back into the cold room.

FIGURE 9.8. Forced air cooling system in operation (Courtesy: Kooljet, Ontario, Canada).

- *Serpentine Cooling:* A serpentine system is designed for bulk bin cooling. It is a modification of the cold-wall method. Bulk bins have vented bottoms with or without side ventilation. Bins are stacked several high and several deep with the forklift openings against the cold wall. Every other forklift opening—sealed with canvas—in the stack matches a cold wall opening. The alternate unsealed forklift openings allow cold air to circulate through the produce. Cold room air is drawn through the produce via the alternate unsealed openings in the stack and the top of the bin.

Hydro-cooling

Hydro-cooling makes use of water as the cooling medium. Water is a better heat transfer medium than air. Values of heat transfer coefficients associated with water are much larger than those associated with air (see Table 9.2). Produce containers or the produce itself is either immersed in a tank or showered with a spray of cold water. To be successful, both the produce and containers must be water tolerant. As water is generally treated with chlorine for sanitation purposes, the produce must be tolerant of low levels of chlorine (50–200 parts per million chlorine). Water sanitation and proper handling are required to prevent spreading of diseases. The advantage is that while cooling water, cooling also can help to cleanse the produce. The first stage of the postharvest process includes washing and can be included as a pre-cooling technique followed by proper rinsing and finish cooling in sanitized water. Since the produce is always in contact with water, there is no cause for moisture loss from the produce. Hence relative humidity of the air is not a consideration in hydro-cooling. However, the packages and containers must be water resistant. Often wax coated containers are used for this purpose. A typical commercial system is illustrated in Figure 9.9.

Hydro-cooling can be carried out either in batch or continuous mode. The treatment again can be given either as an immersion or a spray. Generally, in the continuous system, produce moves through the shower of water and is discharged at the opposite end of the chamber on a conveyor. In the batch system, the produce containers are properly stacked in the chamber and then cold water is showered on the produce or the containers are immersed in a tank containing cold water. The produce is removed when the cooling is complete. The basic equipment of the hydro-cooling system includes: water tank/chamber, pumps, a

FIGURE 9.9. A commercial hydro-cooling system (Courtesy: Kooljet, Ontario, Canada).

water discharge chamber and proper circulation device, water treatment facility (filtration, sanitation, etc. if recycled), and the refrigeration unit. The refrigeration needs generally are very high and often ice banks are used to supplement the mechanical refrigeration. Use of underground water which generally is cooler than municipal water helps to reduce the energy demand.

Package Icing

Package icing generally is used in the fields to cool the produce-filled containers using various forms of ice. Depending on the type of system, the filled in ice could be simply finely crushed, ice flakes or an ice slurry. The slurry, which often is called liquid-ice or slushed-ice is a frequently used form because it can easily be pumped into the container and water drained out to deposit a load of ice into the container. Liquid ice is injected in the container and has better contact with the produce than the other forms. This results in fast cooling (contact heat transfer and melting of ice), and provides high relative humidity. Fairly tight packing can be used, but there must be enough vents in the package to drain the water. The packages must be water tolerant. The ice package containers must be transferred into a cold environment as soon as possible to reduce melting.

Good candidates for package icing are: artichokes, asparagus, beets, broccoli, cantaloupes, carrots, cauliflower, endive, green onions, leafy greens, radishes, spinach, sweet corn and watermelon. However, the following are likely to be damaged by direct exposure to ice: strawberries, blueberries, raspberries, tomatoes, squash, green beans, cucumbers, garlic, okra, bulb onions, romaine lettuce and herbs.

Vacuum Cooling

Vacuum cooling is another technique used to rapidly cool fresh leafy vegetables. The produce to be cooled is placed in a vacuum chamber and a high vacuum is applied. Application of vacuum results in two important functions. It lowers the partial pressure of water vapor in the air, thereby enhancing the water vapor pressure deficit (WVPD). The WVPD is the difference in the partial of pressure of water vapor inside the product which is the saturated water vapor pressure and partial pressure of water vapor in the air, which is reduced to almost zero since the chamber is devoid of air (Table 9.6). Hence, the normal transpiration process, which is relatively slow, is enhanced under the vacuum treatment. More importantly, the vacuum lowers the boiling point of water to below room temperature (Table 9.7). This will allow flashing of moisture from the produce. Since there is no external heat supplied, the heat of vaporization comes from the produce itself resulting in an instantaneous cooling of the produce. The vapor produced is removed from the system through a condenser and hence when the vacuum is broken, the produce will not gain the temperature that it lost during the treatment. Since the application of vacuum and its release can be

TABLE 9.6. Wet Bulb Depression and Lowering of Water Vapour Pressure Deficit Under Transient Vacuum Conditions (saturated water vapour pressure at 30°C is 4.25 kPa).

Temperature (°C)	Pressure (kPa)	Partial Pressure (kPa)	WVPD (kPa)	Wet Bulb (°C)
30	1.0	2.125	2.125	22
	0.5	1.106	25	17
	0.1	0.213	5	12
	0.01	0.021	0.5	<5

TABLE 9.7. Boiling Point Depression Under Vacuum.

Pressure (mmHg)	Boiling Point (°C)
760	100
360	80
55	40
10	10
5	1

accomplished within a short time, the resulting cooling is done very rapidly.

The process was first patented in 1944 and commercialized in 1948. It needs a relatively high capital investment in the form of vacuum pump, condenser and a strong chamber to hold the produce under vacuum. It is ideally suitable for produce with large surface area since it provides a better opportunity to surrender moisture. The basic components of the vacuum cooler are: vacuum chamber (strong), vacuum device (powerful), condenser (efficient). The vacuum chamber usually is a strong vessel with a leak-proof enclosure capable of withstanding at least 1 atmosphere pressure differential. The vacuum device can be either a steam injection or mechanical vapor compression system. The condenser usually is the condensing coils of a refrigeration system.

Vacuum cooling works well with vegetables that have high moisture and a large surface area. It is very energy efficient because only the produce that undergoes moisture loss is cooled. The moisture loss is approximately 1% for every 6°C lowering of temperature. Hence for cooling produce at 30°C, up to 5% moisture can be removed from the produce, which may result in quality loss in leafy vegetables. However, prior to vacuum cooling, the produce can be sprayed with water to compensate for the moisture loss. The added water undergoes preferential evaporation thereby protecting the produce. Because the cost of equipment is high, for economic/cost efficiency, it requires handling large quantities of produce. It also is not suitable for all types of produce especially bulky types like melons, the ones with hard skin like mangoes and papaya and those that do not have a good water loss potential. Hence unless the equipment is mobile and used in a timely, efficient manner, it might turn out to be costly. A commercial vacuum cooling system is shown in Figure 9.10.

FIGURE 9.10. A commercial vacuum cooling system (Courtesy: Kooljet, Ontario, Canada).

PRINCIPLES OF REFRIGERATION

Early refrigeration involved the use of ice—generally referred to as ice-refrigeration. Such techniques were popular in the early part of the twentieth century. Ice was allowed to melt in an insulated chamber that contained food products. For ice to melt, it has to absorb latent heat, which comes from the product to be cooled. Generally, since cold air is heavier than warm air, the ice chest is normally placed above the produce so that the produce is cooled by natural convection.

Today, however, the cooling process is mostly achieved by the use of mechanical refrigeration systems. A refrigeration system allows heat to be absorbed in the cold or cooling chamber at a lower temperature, which is then discarded at a higher temperature (at ambient conditions). This does not mean that heat actually is flowing from a lower temperature to a higher temperature. The transfer of heat is achieved by the flow of a refrigerant which absorbs heat at the lower temperature and releases it at the higher temperature. The absorption of heat in most instances results in a change of state of the refrigerant; the most common change occurs from the liquid state to the gaseous state and is by the absorption of the latent heat of vaporization. Unlike the ice refrigeration system which involves melting, the refrigerant used is generally a substance of low boiling point. For example, the boiling point of Freon-12 is –30°C as compared with 100°C for water.

How can the refrigerant absorb heat at the lower temperature and release it at the higher temperature?

It is done by the selective compression and expansion of the refrigerant which alters its boiling point. In the cooling chamber (*evaporator*), the refrigerant will be at low pressure and hence will have a low boiling point. The temperature of product in the cooling chamber will be higher than the boiling point of the refrigerant. Thus the heat flows from the cooling chamber into the refrigerant (i.e., the refrigerant absorbs the heat). The absorption of latent heat causes the low temperature liquid to boil and turns it into a gas. The emerging gas is then compressed in a *compressor* to increase its pressure. The high pressure elevates the temperature of the gaseous refrigerant and its boiling point to levels much higher than the ambient temperature. Hence, it can be brought back to the liquid state, at high pressure, by a water- or air-cooled *condenser* operating at ambient temperature. The heat now flows out from the refrigerant to the condenser water or air flowing externally. The water for use in condensers can be cooled again through water towers or evaporative coolers. The high pressure liquid, now at ambient temperature, is then allowed to suddenly expand (*expansion valve*) with a subsequent release of pressure which causes the liquid to drop in temperature and at the same time lowers its boiling point.

Vapor Compression Mechanical Refrigeration System

The preceding section, in essence, describes the principle of a vapor compression mechanical refrigeration unit. As shown in the figure below (Figure 9.11), the following are its major components: evaporator, compressor, condenser, receiver and an expansion valve.

As the refrigerant flows through the system, its phase changes from liquid to gas and then back to liquid. Its pressure changes from low to high and back to low. Within the system, therefore, there is a low-pressure side and a high-pressure side and likewise a low-temperature side and a high-temperature side.

Expansion Valve: As the refrigerant arrives at the expansion valve the refrigerant is at a moderate (ambient) temperature and high pressure

156   COOLING OF FRUITS AND VEGETABLES

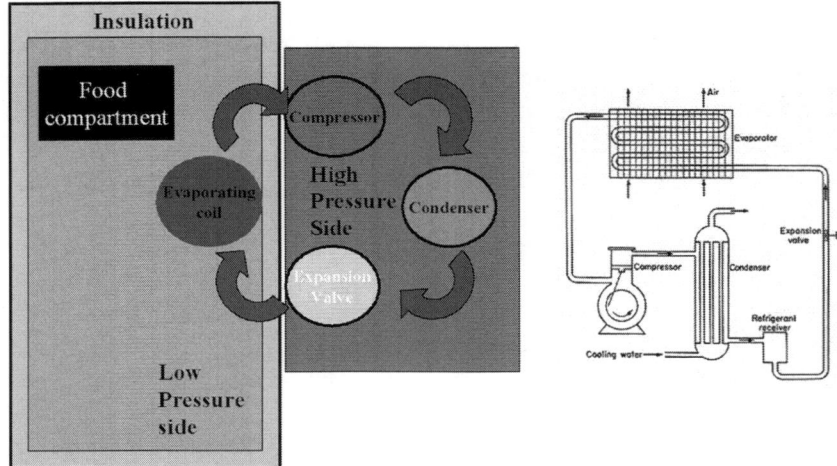

FIGURE 9.11. Schematics of a vapor compression mechanical refrigeration system.

and generally is in a saturated state at or below its condensation temperature. The expansion valve separates the high pressure region from the low pressure region. The expansion valve allows the high pressure liquid to pass through a small opening and allows it to expand and experience a sudden pressure drop. This is accompanied by a concurrent temperature drop (the opposite of what happens in a compressor). Due to the sudden drop in pressure some of the liquid is suddenly converted into a gas. Due to the instantaneous nature of the pressure drop, expansion and vaporization, the liquid appears to "flash-out" from the expansion valve. Various types of expansion valve are available in the market place: (a) manually operated expansion valve, (b) automatic low-side float expansion valve, (c) automatic high-side float expansion valve, (d) automatic expansion valve and (e) thermostatic expansion valve. The manually operated valve will have a hand valve which allows adjustment of an appropriate level of refrigerant flow through the system. In the high- and low-side automatic float systems (b and c), a float controls the flow of refrigerant. When the level falls below a certain critical level, the float activates the flow of liquid refrigerant through the expansion valve. The automatic expansion valve (d) maintains a constant pressure in the evaporator. The increase in pressure in the evaporator causes a diaphragm in the valve to rise against a spring which results in closing of the valve. The valve opens as the pressure drops. The thermostatic expansion valve is perhaps one of the most commonly used valves. It maintains a steady temperature difference between the inlet and exit sections of the evaporator (cooling coils). As the refrigeration need increases, the temperature differential increases, causing more refrigerant to flow through. When the absorption of heat by the refrigerant becomes smaller, the resulting smaller temperature differential reduces the inflow of the refrigerant.

Evaporator: Inside the evaporator, the refrigerant absorbs the heat and vaporizes to a gaseous state. Both direct expansion and indirect expansion evaporators are used commonly. In the former, the cooling coils are in direct contact with the object or liquid being cooled. Indirect evaporators involve the use of a carrier medium such as water or air which is cooled by the refrigerant vaporizing in the cooling coils. This cooled carrier liquid is then used to cool the product. The evaporators can be either bare-pipe, finned-tube or plate type. The bare-pipe types are the most common. Fins added to the pipes provide additional surface area for evaporation of the refrigerant. Plate types usually are used in indirect, i.e., plate heat exchangers.

Compressor: The refrigerant enters the compressor in the vapor state at low pressure from the evaporator. The work done in the compressor raises its pressure and increases its temperature. The high pressure helps in elevating the boiling point of the refrigerant so that it can be cooled into a liquid at ambient temperature in the condenser. Three types of compressors are common: reciprocating, centrifugal and rotary. The reciprocating type contains a piston that travels back and forth in a cylinder (single or multi). These are the most commonly used. The centrifugal type contains an impeller with several blades that turn at high speed. The rotary compressor involves a vane that rotates inside a cylinder.

Condenser: The function of a condenser in the refrigeration system is to transfer the heat from the refrigerant to another medium such as air or water. This will cause the high pressure refrigerant to condense into a liquid. The major types of condensers are: (a) water-cooled, (b) air-cooled and (c) evaporative. Several configurations exist: shell and tube, tube and fin, plate heat exchanger, etc. Domestic refrigerators mostly use air cooled tube-and fin type condensers, while large-scale operations generally involve water-cooled condensers attached to cooling towers operated on evaporative cooling techniques.

Receiver: A receiver is a reservoir for storage of the condensed liquid refrigerant. It helps to stabilize the system by minimizing temperature fluctuations.

Refrigerants

A wide variety of refrigerants are available commercially for use in vapor-compression systems. The selection of a refrigerant is based on several performance characteristics. The following parameters often are considered:

1. *Latent heat of vaporization:* Higher the better.
2. *Condensing pressure:* Moderate ones are better.
3. *Freezing point:* Should be below the evaporator temperature.
4. *Critical temperature:* This should be high. Above the critical temperature, the refrigerant cannot be liquefied. Hence this should be sufficiently above the highest condenser (ambient) temperature.
5. *Toxicity:* Should be low, preferably non toxic.
6. *Flammability:* Should be inflammable.
7. *Corrosiveness:* Should be less corrosive on metal surfaces.
8. *Chemical stability:* Should be stable.
9. *Detection of leaks:* Any developed leaks should be easily detectable.
10. *Cost:* Low cost desirable.
11. *Environmental impact:* The refrigerant released by the system due to leaks should not cause environmental damage.

Ammonia was one of the earliest refrigerants used and offers an exceptionally high latent heat of vaporization. It is non-corrosive to iron and steel, but corrodes copper, brass and bronze. It is irritating to mucous membranes and eyes. It can be toxic at concentrations above 0.5% level. The leak can easily be detected by smell or burning sulfur candles (causes white smoke). Ammonia leaks also have been known to damage produce. Hence, it is not commonly used in storage systems.

Most of the refrigerants used today in commercial systems are halocarbons. Freon-12, also called Refrigerant-12 or R-12 is one of the most commonly used refrigerants in air conditioning systems. Freon is the trade name for the refrigerant manufactured by Dupont and chemically it is a dichlorodifluoromethane (CCl_2F_2). R-22 is chlorodifluoromethane ($CHClF_2$). This refrigerant has been particularly useful in systems involving very low temperatures (–40 to –70°C). R-30 is methylenechloride (CH_2Cl_2).

During the 1970s, it was postulated that chlorofluorocarbons (CFCs), because of their extremely stable characteristics, have a long life in the lower atmosphere and gradually migrate upward over a period of time. Here, the chlorine portion of the CFC splits due to exposure to the sun's UV rays and reacts with the ozone resulting in the depletion of ozone concentration. With depletion of the protective ozone layer surrounding the earth, we lose protection from the sun's harmful UV rays. Many of the commonly used refrigerants are fully halogenated chlorofluorocarbons containing chlorine.

Alternatives to CFCs are being actively considered as replacement for these refrigerants. The hydrofluorocarbons (HFC) and hydrochlorofluorocarbons (HCFCs) have been increasingly studied as alternatives. Hydrogen-containing fluorocarbons have weak carhon-hydrogen bonds that are more susceptible to cleavage and are postulated to be less stable than conventional CFCs.

Vapor Absorption Refrigeration System

The vapor absorption refrigeration system generally is referred to as a "gas refrigeration system" because it was developed for burning gas to provide the external energy for moving heat from a low-temperature region to an external high temperature region. This heat also can be provided by other means, such as the burning of coal, fuel oil or electricity or solar heating. The absorption refrigeration system consists of a generator, evaporator, absorber and condenser, all connected in a closed system (see figure Figure 9.12). Conceptually, an absorption-refrigeration machine is similar to a vapor-compression system in which the compressor is substituted by four elements: a vapour absorber based on another liquid, a pump for the liquid solution, a generator or boiler to release the vapour from solution, and a valve to recycle the absorbent liquid.

One of the big advantages of the system is that its cycle requires less work to operate (only that of the pump), or none at all if the liquid is naturally pumped by gravity in a thermo-siphon, at the expense of an additional heat source required at the regenerator. As with a mechanical vapor compression system, cooling water is required to carry the heat away from the refrigerant. The system operates also on liquid-gas states, as does the compression system, but the total pressure is the same throughout the system. The movement of the refrigerant is facilitated by a carrying medium which absorbs the refrigerant in the evaporator and releases it to the condenser.

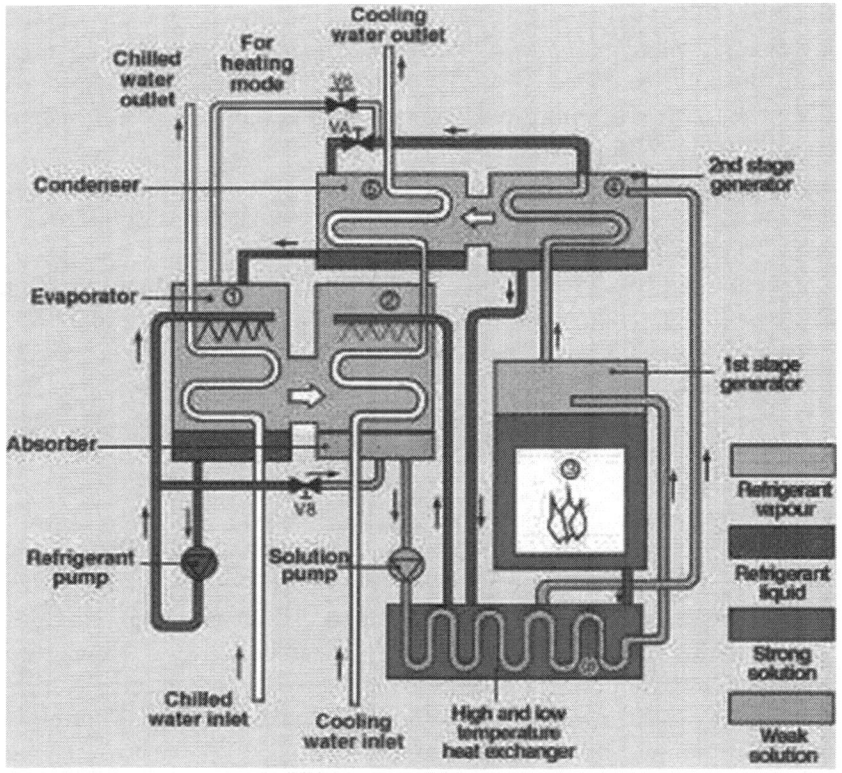

FIGURE 9.12. *Vapor absorption refrigeration system (Courtesy: Voltas Corp.).*

Absorber—Absorbent: An absorbent is used to absorb the refrigerant and carry it from the low refrigerant vapor pressure to the high vapor pressure side. The absorbent also carries the heat to the absorber which is then removed from the system.

Ammonia—water (where ammonia is the refrigerant and water is the absorbent) is the most common vapor absorbent refrigeration system. In this system, ammonia is given off as the ammonia—water solution is heated. Nearly all of the ammonia is released at the boiling point of water. As ammonia evaporates, it absorbs the latent heat from the water vapor which then condenses and flows back to the absorber. In the absorber the water in the system is cooled by water external to the refrigeration system. The cool water within the system can absorb ammonia (solubility of gas increases at lower temperatures) as it passes through the absorber and returns to the generator.

Generator: The generator contains ammonia dissolved in water. Heat is provided from an external source to drive off the ammonia vapor and to return the solution to the generator from the absorber. The generator serves the same function in the absorption system as the compressor in the compression system.

Condenser: The condenser receives the ammonia vapor (driven off from solution via generator) and condenses it to a liquid. Water is furnished externally to carry the heat from the gas (in the same way water is supplied to the condenser of a compression system). The water then goes to the external surfaces of the absorber for removing additional heat.

Evaporator: The ammonia liquid enters the evaporator or cooling coils. Hydrogen gas (moderator) from the absorber also enters the evaporator. The cool ammonia absorbs heat in the evaporator and returns to the gaseous state, and moves to the absorber. The ammonia concentration in the absorber is lower due to removal of ammonia from the gas mixture. The hydrogen present in the system migrates as necessary to maintain the system pressure constant (the gas partial pressure varies at different locations around the system, but the total system pressure remains constant). Ammonia gas and hydrogen move from the evaporator into the absorber. Water falling down through the absorber dissolves ammonia and thus lowers the ammonia vapor pressure there. This draws more ammonia into the absorber. The ammonia water solution then moves to the generator where it can be heated as necessary.

The three basic components of the absorption system are: ammonia moves from the generator to the evaporator, to the absorber, and back to the generator. Water moves from the generator to the absorber and back to the generator. Hydrogen moves from the absorber to the evaporator and back to the absorber. There are three gas cycles imposed upon each other to furnish the desired heat transfer process. In normal operating conditions, the condenser pressure will be about 12 atm (180 psi) and the absorber pressure will be 1.5 atm (25 psi) with respect to ammonia.

Evaporative Cooling

Evaporative cooling is based on the principles of transpiration which were discussed previously. Mixing water and non-saturated air produces a refrigerant effect (i.e., a temperature drop below ambient temperature). This is an old technique—the one used by ancient Egyptians to cool drinking water in porous earthen pots, and to cool space by splashing some water on the floor, and even to produce ice in deserts under carefully controlled conditions.

The refrigeration effect comes from the energy demanded by evaporating water (equal to the vaporisation enthalpy), a natural process driven by air dryness. Closely related to evaporative cooling is vacuum cooling when a vacuum, previously discussed, is applied to a liquid or solid (usually in aqueous solutions). Evaporative cooling is not generally considered a powerful source of refrigeration since it is rather limited in practice to only a slight cooling of the water or the air that is fed to the system. The other handicap is that evaporation is a slow process; the air becomes saturated during the process and must be dehumidified to reuse for the purpose of evaporative cooling. However, new developments in desiccant regeneration are showing promise, particularly for air-conditioning applications. Further, in some cold storage applications cool humidified air is really not a disadvantage at all.

An evaporative cooling system schematic is shown below in Figure 9.13. The cooler consists of a ceramic bed of perforated material through which water is made to trickle down. Air to be cooled is forced up through the bed and meets the incoming water. Evaporation results due to heat transfer, and the air temperature falls along the wet bulb line. The air is simultaneously humidified. Cool humidified air coming off the cooler can then be used in cold storage or for cooling of pro-

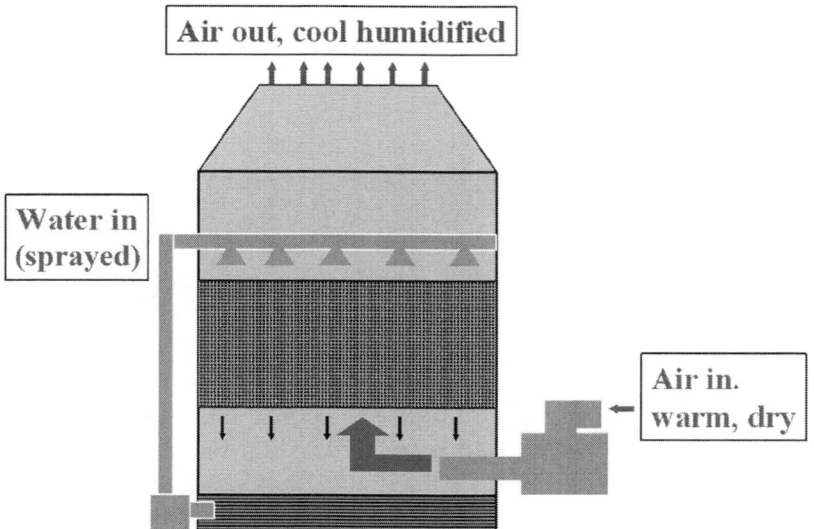

FIGURE 9.13. Schematics of the evaporative cooling system.

duce. A dry inlet air will result in the maximum wet-bulb depression. The residence time of air in the bed is generally adjusted to accomplish complete saturation of the exiting air.

Night Cooling

Cooling at night time is applicable only when there is significant temperature difference between day and night temperatures. There are many places and times when such differences do exist and night temperatures can be 15–20°C lower than the daytime maximum temperatures. Under such situations, the stored commodity is thoroughly ventilated during night times in ventilated storage systems, which generally do not depend on external refrigeration. Generally, during the day, only recirculation of indoor air is employed. Secondly, this situation can be advantageous by harvesting the produce during the night, which can significantly reduce the pre-cooling load.

High-Altitude Cooling

It generally is recognized that air temperature decreases 10°C for every 1000m increase in altitude. Therefore, the temperatures at high elevations generally are lower than in the valleys. Can this be used for produce cooling? Is it possible to bring down cold air from high altitudes and use it in cold storage in the valleys where fruits and vegetables are grown? It is a good question. Of course, it is possible to erect large ducts running 1000s of meters (however expensive) and to draw air down from the mountains, but not the cold air. The air on the top of a mountain is cooler because it has gone through adiabatic expansion. The density of such air is lower. When the air is forced down, the adiabatic compression that occurs due to the pressure of the air column results in warming the temperature to ambient conditions if not more, because of the added work energy. However, it is certainly possible to build cold storage at higher altitudes and to move the produce to those facilities to take advantage of the cooler conditions.

Radiant Cooling

Radiant cooling is the reverse of radiant heating. Solar collectors are used to trap solar energy for various heating applications, including heating of water used in kitchens and bathrooms. While these will trap the sun's energy during the day, the same process can be used to lower the air temperature by connecting them to the ventilation system of the building. At night the outside temperature will be lower than inside and the solar collector will help to disperse the heat to the environment. This is not very efficient for large-scale cooling, but temperatures can be lowered by about 4–5°C below the night time outside temperature.

Well Water and Underground Structures

One can also use well water for cooling applications. Well water often is much cooler than air and generally its temperature stays relatively constant. Water can be lifted from deep wells or bore-wells and used for cooling, as in hydro-cooling operations. Since underground temperatures are cooler, cold storage can be constructed in underground structures. One can also use naturally occurring ice in winter climates in cold countries. These concepts, however, are rarely used in practice.

Transit Cooling

Package icing, use of mechanical refrigeration or the use of cryogenics are techniques normally employed for transit cooling. Package icing was discussed earlier as one of the field precooling techniques. In cryogenic cooling, liquid nitrogen and carbon dioxide are frequently employed. Liquid nitrogen is a low boiling pressurized liquid that has a boiling point of around −196°C. It is commercially available in pressurized cylinders. It is environmentally friendly since natural air contains almost 80% nitrogen. It can be used as a thermostatically controlled spray in transportation facilities like trucks or rail cars or marine containers. The liquid boils off as soon as it is released to atmospheric conditions and can be properly circulated using a fan and a well-configured distribution system. It requires very minimal equipment and is currently widely used in commercial transportation facilities. Only a temperature that is too low can be occasionally problematic, but with proper thermostatic control it can be kept at a proper level.

Carbon dioxide can be used either as a liquid or solid (dry ice). It adds CO_2 to the environment. This can help up to a certain level but excess must be vented to prevent undesirable effects on produce quality. It is used often more in freezing equipment than for cooling.

Thermoelectric Cooling

Thermoelectic cooling is based on a reverse Seebeck effect. When the two junctions of dissimilar metals are placed at different temperatures, an electromagnetic field (emf) is generated that flows through the circuit. The generated emf is proportional to the temperature difference. When one is known, the other can be estimated. Thermocouple thermometry is based on this approach. Solid-state electrically-driven refrigerators (also named thermo-electric coolers, TEC) are based on the Peltier effect. When a DC current flow in a circuit is formed by two dissimilar electrical conductors, heat is absorbed at one junction and more heat is released at the other junction, reversing the effects when reversing the direction of the current. A typical thermoelectric cooler consists of pairs of p-type and n-type semiconductor thermo-elements forming thermocouple junctions that are connected electrically in series and thermally in parallel. A simplified thermoelectric cooling concept is illustrated in Figure 9.14. When the cooler is connected to a DC power source, the cold side of the module will cool down until the internal heat conduction balances the heat-pump capability—e.g., starting at room

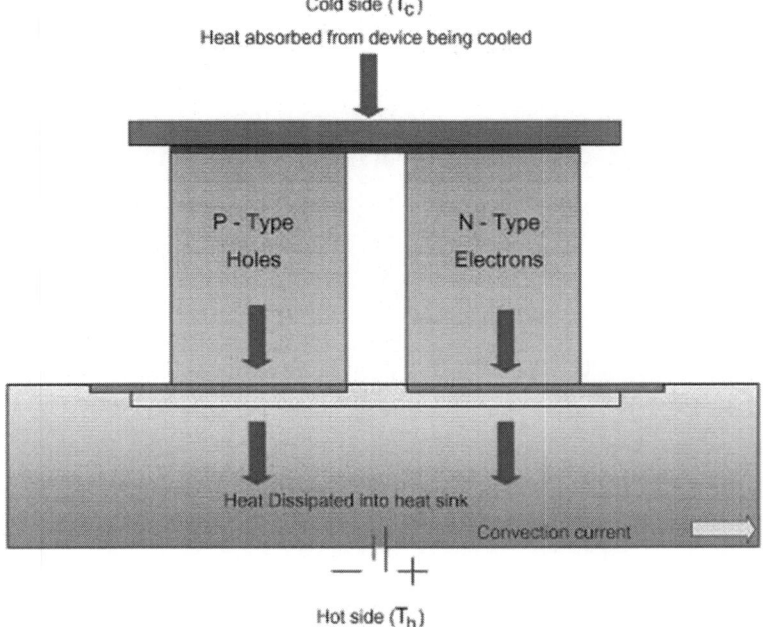

FIGURE 9.14. Schematic illustration the thermoelectric cooler.

166 COOLING OF FRUITS AND VEGETABLES

FIGURE 9.15. A commercially available thermo-electric cooler.

temperature of 20°C, a steady state may be reached with the cold side at –40°C if well insulated, and the hot side at say 30°C if fan-cooled or near 20°C if vigorously cooled. Figure 9.15 illustrates a commercially available thermo-electric cooler that is commonly available.

CHAPTER 10

Cold Storage Systems for Fruits and Vegetables

INTRODUCTION

A cold storage system generally is referred to as a storage "system" because of the need to control several components: temperature, relative humidity (RH), air velocity, air composition, etc. Proper control depends on structural and constructional details, insulation, air tightness and vapor permeability. The degree of control depends on the type of storage facility and needed functions.

While simple storage systems, such as ventilated storage for bulk storage of potatoes or dry onions, only take into account the thermal insulation, air circulation and fan assisted ventilation, others rely on more elaborate controls. Even standard regular atmosphere cold storage needs to take into account the appropriately designed refrigeration equipment to meet the needed goal in addition to proper humidity and air distribution controls. In more complex storage systems such as CA and hypobaric storage, thermal insulation, air tightness, vapor barrier control, precise control of temperature, RH and air composition need to be taken into account.

Temperature

Temperature is the key factor in postharvest storage. Produce must be stored at the lowest permissible temperature. Some tropical produce could be chilling sensitive, and these should not be stored at temperatures below which they are sensitive to cold injury (generally the lower limit is around 10°C). Each produce has its own optimal storage temperature, which generally is the lowest that causes no chilling or freezing injury.

Table 10.1 lists typical cooling and cold storage conditions for fruits and vegetables. This can be used as a guideline. Every degree above the best temperature reduces the storage life. The Q10 concept discussed in Chapter 7 on respiration can be used with respiration rates, heat generation rates or quality deterioration rates. It can be used to predict the produce quality following storage.

Ideally, the produce must be stored at its optimum condition, which might mean a separate storage room for each commodity, which is not practical. Mixed loads are more common and practical. With mixed loads, product temperature compatibility must be checked. Generally, produce with widely differing temperature tolerances should not be mixed. This may result in quality loss in some produce while being beneficial for others. This means chilling-sensitive commodities can be grouped together in rooms maintained around 10°C while the non-chilling-sensitive commodities can be stored as low as 2°C. Compromise should be based on value: Don't store the sensitive product in the same place long enough to produce damage. Compatibility also must be considered with respect to ethylene tolerance. Produce that does not tolerate ethylene should not be stored along with that which generates too much ethylene (see chapter on classification for details). Compatibility with respect to relative humidity is next. Some vegetables can tolerate (and in fact prefer) slightly lower humidity conditions like onions and potatoes, while leafy produce requires almost saturated humidity conditions. Hence, it is preferable not to store them in the same room. Odor compatibility is the next. Some fruits and vegetables can pick up odors; don't store fish in the same room as fruits and vegetables.

The optimum storage temperature for most temperate produce is 30–32°F for the non-chilling-sensitive varieties and 38–40°F for the chilling-sensitive varieties. For tropical produce the desirable temperature ranges are: 45–50°F (non-chilling); 50–60°F (chilling-sensitive).

From a storage point of view:

1. Temperature fluctuation in a storage chamber should be within ±2°F. Larger temperature fluctuations will result in a loss of produce quality. Warm produce coming in contact with cold air will result in product desiccation if the air is not saturated and result

TABLE 10.1. Recommended Storage Temperatures for Fruits and Vegetables.

Product	Temperature °C	Temperature °F	Relative Humidity (percent)	Approximate Storage Life
Amaranth	0–2	32–36	95–100	10–14 days
Anise	0–2	32–36	90–95	2–3 weeks
Apples	–1–4	30–40	90–95	1–12 months
Apricots	–0.5–0	31–32	90–95	1–3 weeks
Artichokes, globe	0	32	95–100	2–3 weeks
Asian pear	1	34	90–95	5–6 months
Asparagus	0–2	32–35	95–100	2–3 weeks
Atemoya	13	55	85–90	4–6 weeks
Avocados, Fuerte, Hass	7	45	85–90	2 weeks
Avocados, Lula, Booth-1	4	40	90–95	4–8 weeks
Avocados, Fuchs, Pollock	13	55	85–90	2 weeks
Babaco	7	45	85–90	1–3 weeks
Bananas, green	13–14	56–58	90–95	14 weeks
Barbados cherry	0	32	85–90	7–8 weeks
Bean sprouts	0	32	95–100	7–9 days
Beans, dry	4–10	40–50	40–50	6–10 months
Beans, green or snap	4–7	40–45	95	7–10 days
Beans, lima, in pods	5–6	41–43	95	5 days
Beets, bunched	0	32	98–100	10–14 days
Beets, topped	0	32	98–100	4–6 months
Belgian endive	2–3	36–38	95–98	24 weeks
Bitter melon	12–13	53–55	85–90	2–3 weeks
Black sapote	13–15	55–60	85–90	2–3 weeks
Blackberries	–0.5–0	31–32	90–95	2–3 days
Blood orange	4–7	40–44	90–95	3–8 weeks
Blueberries	–0.5–0	31–32	90–95	2 weeks
Bok choy	0	32	95–100	3 weeks
Boniato	13–15	55–60	85–90	4–5 months
Breadfruit	13–15	55–60	85–90	2–6 weeks
Broccoli	0	32	95–100	10–14 days
Brussels sprouts	0	32	95–100	3–5 weeks
Cabbage, early	0	32	98–100	3–6 weeks
Cabbage, late	0	32	98–100	5–6 months
Cactus Leaves	24	36–40	90–95	3 weeks
Cactus Pear	24	36–40	90–95	3 weeks
Caimito	3	38	90	3 weeks
Calabaza	10–13	50–55	50–70	2–3 months
Calamondin	9–10	48–50	90	2 weeks
Canistel	13–15	55–60	85–90	3 weeks

(continued)

TABLE 10.1 (continued). Recommended Storage Temperatures for Fruits and Vegetables.

Product	Temperature °C	°F	Relative Humidity (percent)	Approximate Storage Life
Cantaloups (3/4–slip)	2–5	36–41	95	15 days
Cantaloups (full-slip)	0–2	32–36	95	5–14 days
Carambola	9–10	48–50	85–90	3–4 weeks
Carrots, bunched	0	32	95–100	2 weeks
Carrots, mature	0	32	98–100	7–9 months
Carrots, immature	0	32	98–100	4–6 weeks
Cashew apple	0–2	32–36	95–98	5 weeks
Cauliflower	0	32	85–90	34 weeks
Celeriac	0	32	97–99	6–8 months
Celery	0	32	98–100	2–3 months
Chard	0	32	95–100	10–14 days
Chayote squash	7	45	85–90	4–6 weeks
Cherimoya	13	55	90–95	2–4 weeks
Cherries, sour	0	32	90–95	3–7 days
Cherries, sweet	−1 to −0.5	30–31	90–95	2–3 weeks
Chinese broccoli	0	32	95–100	10–14 days
Chinese cabbage	0	32	95–100	2–3 months
Chinese long bean	4–7	40–45	90–95	7–10 days
Clementine	4	40	90–95	24 weeks
Coconuts	0–1.5	32–35	80–85	1–2 months
Collards	0	32	95–100	10–14 days
Corn, sweet	0	32	95–98	5–8 days
Cranberries	2–4	36–40	90–95	24 months
Cucumbers	10–13	50–55	95	10–14 days
Currants	−0.5–0	31–32	90–95	1–4 weeks
Custard apples	5–7	41–45	85–90	4–6 weeks
Daikon	0–1	32–34	95–100	4 months
Dates	−18 or 0	0 or 32	75	6–12 months
Dewberries	−0.5–0	31–32	90–95	2–3 days
Durian	4–6	39–42	85–90	6–8 weeks
Eggplants	12	54	90–95	1 week
Elderberries	−0.5–0	31–32	90–95	1–2 weeks
Endive and escarole	0	32	95–100	2–3 weeks
Feijoa	5–10	41–50	90	2–3 weeks
Figs fresh	−0.5–0	31–32	85–90	7–10 days
Garlic	0	32	65–70	6–7 months
Ginger root	13	55	65	6 months
Gooseberries	−0.5–0	31–32	90–95	34 weeks
Granadilla	10	50	85–90	3–4 weeks

(continued)

TABLE 10.1 (continued). Recommended Storage Temperatures for Fruits and Vegetables.

Product	Temperature °C	°F	Relative Humidity (percent)	Approximate Storage Life
Grapefruit, Calif. & Ariz.	14–15	58–60	85–90	6–8 weeks
Grapefruit, Fla. & Texas	10–15	50–60	85–90	6–8 weeks
Grapes, Vinifera	–1 to –0.5	30–31	90–95	1–6 months
Grapes, American	–0.5–0	31–32	85	2–8 weeks
Greens, leafy	0	32	95–100	10–14 days
Guavas	5–10	41–50	90	2–3 weeks
Haricot vert	4–7	40–45	95	7–10 days
Horseradish	–1–0	30–32	98–100	10–12 months
Jaboticaba	13–15	55–60	90–95	2–3 days
Jackfruit	13	55	85–90	2–6 weeks
Jaffa orange	8–10	46–50	85–90	8–12 weeks
Japanese eggplant	8–12	46–54	90–95	1 week
Jerusalem Artichoke	–0.5–0	31–32	90–95	+5 months
Jicama	13–18	55–65	65–70	1–2 months
Kale	0	32	95–100	2–3 weeks
Kiwano	10–15	50–60	90	6 months
Kiwifruit	0	32	90–95	3–5 months
Kohlrabi	0	32	98–100	2–3 months
Kumquats	4	40	90–95	2–4 weeks
Langsat	11–14	52–58	85–90	2 weeks
Leeks	0	32	95–100	2–3 months
Lemons	10–13	50–55	85–90	1–6 months
Lettuce	0	32	98–100	2–3 weeks
Limes	9–10	48–50	85–90	6–8 weeks
Lo bok	0–1.5	32–35	95–100	24 months
Loganberries	–0.5–0	31–32	90–95	2–3 days
Longan	1.5	35	90–95	3–5 weeks
Loquats	0	32	90	3 weeks
Lychees	1.5	35	90–95	3–5 weeks
Malanga	7	45	70–80	3 months
Mamey	13–15	55–60	90–95	2–6 weeks
Mangoes	13	55	85–90	2–3 weeks
Mangosteen	13	55	85–90	2–4 weeks
Melon, Casaba	10	50	90–95	3 weeks
Melon, Crenshaw	7	45	90–95	2 weeks
Melon, Honeydew	7	45	90–95	3 weeks
Melon, Persian	7	45	90–95	2 weeks
Mushrooms	0	32	95	34 days
Nectarines	–0.5–0	31–32	90–95	2–4 weeks

(continued)

TABLE 10.1 (continued). Recommended Storage Temperatures for Fruits and Vegetables.

Product	Temperature °C	°F	Relative Humidity (percent)	Approximate Storage Life
Okra	7–10	45–50	90–95	7–10 days
Olives, fresh	5–10	41–50	85–90	+6 weeks
Onions, green	0	32	95–100	3–4 weeks
Onions, dry	0	32	65–70	1–8 months
Onion sets	0	32	65–70	6–8 months
Oranges, Calif. & Ariz.	3–9	38–48	85–90	3–8 weeks
Oranges, Fla. & Texas	0–1	32–34	85–90	8–12 weeks
Papayas	7–13	45–55	85–90	1–3 weeks
Passionfruit	7–10	45–50	85–90	3–5 weeks
Parsley	0	32	95–100	2–2.5 months
Parsnips	0	32	95–100	+6 months
Peaches	−0.5–0	31–32	90–95	2–4 weeks
Pears	−1.5 to −0.5	29–31	90–95	2–7 months
Peas, green	0	32	95–98	1–2 weeks
Peas, southern	+5	40–41	95	6–8 days
Pepino	4	40	85–90	1 month
Peppers, Chili (dry)	0–10	32–50	60–70	6 months
Peppers, sweet	7–13	45–55	90–95	2–3 weeks
Persimmons, Japanese	−1	30	90	3–4 months
Pineapples	7–13	45–55	85–90	2–4 weeks
Plantain	13–14	55–58	90–95	1–5 weeks
Plums and prunes	−0.5–0	31–32	90–95	2–5 weeks
Pomegranates	5	41	90–95	2–3 months
Potatoes, early crop	10–16	50–60	90–95	10–14 days
Potatoes, late crop	4.5–13	40–55	90–95	5–10 months
Pummelo	7–9	45–48	85–90	12 weeks
Pumpkins	10–13	50–55	50–70	2–3 months
Quinces	−0.5–0	31–32	90	2–3 months
Raddichio	0–1	32–34	95–100	2–3 weeks
Radishes, spring	0	32	95–100	3–4 weeks
Radishes, winter	0	32	95–100	2–4 months
Rambutan	12	54	90–95	1–3 weeks
Raspberries	−0.5–0	31–32	90–95	2–3 days
Rhubarb	0	32	95–100	2–4 weeks
Rutabagas	0	32	98–100	+6 months
Salsify	0	32	95–98	2–4 months
Santol	7–9	45–48	85–90	3 weeks
Sapodilla	16–20	60–68	85–90	2–3 weeks
Scorzonera	0–1	32–34	95–98	6 months

(continued)

TABLE 10.1 (continued). Recommended Storage Temperatures for Fruits and Vegetables.

Product	Temperature °C	Temperature °F	Relative Humidity (percent)	Approximate Storage Life
Seedless cucumbers	10–13	50–55	85–90	10–14 days
Snow peas	0–1	32–34	90–95	1–2 weeks
Soursop	13	55	85–90	1–2 weeks
Spinach	0	32	95–100	10–14 days
Squashes, summer	5–10	41–50	95	1–2 weeks
Squashes, winter	10	50	50–70	2–3 months
Strawberries	0	32	90–95	5–7 days
Sugar apples	7	45	85–90	4 weeks
Sweet potatoes	13–15	55–60	85–90	4–7 months
Tamarillos	3–4	37–40	85–95	10 weeks
Tamarinds	7	45	90–95	3–4 weeks
Tangerines, mandarins, and related citrus fruits	4	40	90–95	24 weeks
Taro root	7–10	45–50	85–90	4–5 months
Tomatillos	13–15	55–60	85–90	1–3 weeks
Tomatoes, mature–green	18–22	65–72	90–95	1–3 weeks
Tomatoes, firm–ripe	13–15	55–60	90–95	4–7 days
Turnips	0	32	95	4–5 months
Turnip greens	0	32	95–100	10–14 days
Ugli fruit	4	40	90–95	2–3 weeks
Water chestnuts	0–2	32–36	98–100	1–2 months
Watercress	0	32	95–100	2–3 weeks
Watermelons	10–15	50–60	90	2–3 weeks
White sapote	19–21	67–70	85–90	2–3 weeks
White asparagus	0–2	32–36	95–100	2–3 weeks
Winged bean	10	50	90	4 weeks
Yams	16	61	70–80	6–7 months
Yucca root	0–5	32–41	85–90	1–2 months

Source: FAO http://www.fao.org/wairdocs/X5403E/x5403e09.htm (McGregor, B.M. 1989. Tropical Products Transport Handbook. USDA Office of Transportation, Agricultural Handbook 668).

in produce sweating if the air is saturated. On the other hand, cold produce coming in contact with warm air will result in desiccation if the WVPD is positive or condensation of water on produce surface if WVPD is negative. Moisture from the desiccated or sweated produce eventually is carried away by air to the cooling coils. The larger the temperature fluctuation the greater is the loss.

2. Adequate refrigeration capacity to handle loads: The designed refrigeration capacity should take into account normal load as well as unexpected loads. Storage life of existing and incoming commodities is maximized by pre-cooling the incoming produce to the storage temperature. On the other hand, storage life of all commodities will be at risk with warm products coming into the storage, especially when refrigeration capacity is limited. Undersized refrigeration capacity will result in slow cooling and can cause product warming and large temperature fluctuations. Oversized refrigeration capacity is more expensive and involves more capital investment. Several small size units acting in parallel are preferred over one large one, as they will cause less temperature fluctuation in the chamber.
3. Room temperature should be close to the refrigeration temperature. This causes less temperature fluctuation when the control system goes on and off between the set limits.
4. The refrigeration system and cooling coils should have a large heat transfer surface area. This will result in more efficient heat transfer; fins can be used to increase surface area.
5. Good air velocity increases proper temperature distribution. Generally, air flow rates of 50–75 cubic feet per minute are desirable; higher flow rates will cause desiccation, especially when a large temperature difference exists between coil and produce; lower flow rates will lead to temperature stratification and improper cooling of the produce.
6. The room should have adequate thermal insulation to prevent heat ingress from walls, doors, floor, ceiling, etc.
7. The room should have an appropriate temperature control system, and the temperature should be measured at appropriate locations—preferably not near door or coils.

Relative Humidity

Relative humidity aspect was discussed in detail with reference to transpiration. Try to trace the air path in a storage chamber on a psychrometric chart. Assume that the room is designed for storage of tropical produce at 15°C, 90% RH. Assume an overall temperature fluctuation of +5°C with the cooling coils operating at 10°C (See Figure 10.1). This means that air coming through the produce stack could warm up to 20°C with an assumed RH of 80%. Notice that this is not an adia-

Introduction

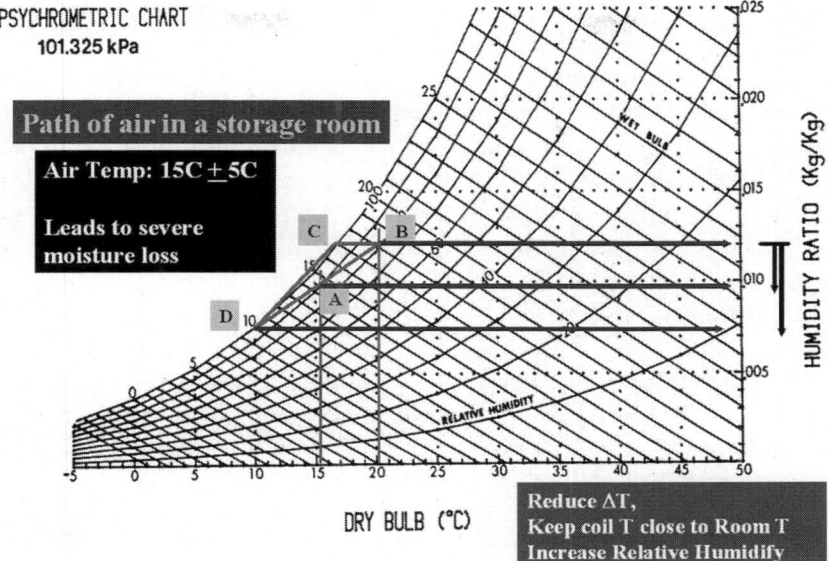

FIGURE 10.1. Psychrometric representation of air movement in a storage room.

batic saturation process because the air picks up heat of respiration, and heat load from the room in addition to heat of vaporization. The warm air travels to the cooling coil where its temperature is lowered to 10°C; moisture condenses on the coil as the air temperature falls below 17°C (dew point temperature of 20°C and 80% RH air). The air leaving the cooling coil will be saturated at 10°C; however, as it travels through the stack it warms up to 20°C and 80% RH. The difference between these two humidity values (0.012–0.0075 = 0.045 kg H_2O per kg dry air) is the amount of moisture picked up by the air through each cycle. In the cyclic process of air traveling around, moisture is continuously removed from the produce.

Moisture loss can be reduced by narrowing down the temperature difference between the coil and the air. Moisture loss can also be reduced by narrowing down the temperature fluctuation as well. Moisture loss is offset by humidifying the air externally using special techniques for storage: a jacketed storage system, Filacell® storage system or fog-spray system.

Air Velocity

Air circulation can be natural or forced, the latter being common in storage systems. Air circulation promotes both heat and mass transfer; while heat transfer is desirable, mass transfer is not. Air velocity is important for establishing less temperature fluctuation, also for bringing down the temperature difference between air and cooling coil. Further, air circulation also is important for establishing uniformity of CA conditions. The air circulation pattern depends on the nature of circulation (natural or forced), fan type and capacity, the air delivery system, as well as produce package design and stacking.

Atmospheric Composition

In regular atmosphere systems, the atmospheric composition is not controlled. Excessive CO_2 & ethylene accumulation is prevented by scrubbing. In CA/MA storage systems, the atmosphere is altered. Low oxygen levels and high CO_2 levels are intentionally created to reduce the rate of respiration and heat production. Removal of ethylene (C_2H_4) is a must in these systems because ethylene even in trace quantities can trigger accelerated ripening of some fruits. Sometimes, carbon monoxide (CO) is added as a CA supplement. It has been shown to inhibit discoloration of cut surfaces. It also has been shown to have some fungistatic activity at a 5–10% concentration level. It is flammable at > 12.5% concentration. Addition of 2–3% CO at 2% O_2 has been found to be successful for lettuce.

Structural Aspects

A fruit and vegetable storage chamber is built to provide storage at a desired temperature that is normally different from the ambient (room temperature could be cooler than ambient in summer and warmer than ambient in winter months) with some control over the relative humidity and air flow rate. Additionally, specialized storage chambers are built that permit controlled modifications of air composition to desired levels (for example 2.5% oxygen and 5% carbon dioxide) different from those at which they normally exist. While most construction would require appropriate thermal insulation and structural rigidity, the specialized ones need to take into account vapor and gas barrier properties as well. The following considers structural details in the construction of these storages.

1. Building Materials
    a. Concrete: durable, noncombustible, waterproof, pest proof;

b. Masonry blocks—concrete blocks bonded by cement mortar: popular but porous, not vapor and gas proof;
c. Wood frame with insulating materials;
d. Steel frame with insulating materials.
2. Common insulating materials: fiberglass, expanded polystyrene, polyurethane, urea formaldehyde resins.
3. Water vapor barriers and gas seals: sprayed on polyurethane, good vapor and gas barrier.
4. Accessibility to utilities, transportation.
5. Size planning: single large or multiple small, etc.
6. Appropriate instrumentation: Important especially with the more sophisticated storage techniques. Instrumentation required for both monitoring and control.
   a. Temperature: thermometry.
   b. Relative humidity: hygrometers (sling, hair).
   c. Air velocity: anemometers (vane or hot wire).
   d. Gas composition: Gas chromatography.

REFRIGERATION REQUIREMENTS

In order to calculate the refrigeration requirements of any storage system, several factors must be considered. Although cold storage systems are intended for storing pre-cooled produce, the calculations also should take into account loading of warm product when that happens frequently.

The following are the major factors contributing to the refrigeration load:

1. Building transmission load: Refers to heat gain through the building walls, ceiling and floor.
2. Air exchange load: Contributed by opening of the doors and ventilation.
3. Product load:
   a. This is contributed by product cooling (removal of sensible heat at the appropriate rate of cooling), if applicable, and
   b. Heat input from the respiration of produce (undergoing pre-cooling and stored).

4. Miscellaneous heat loads: This includes heat input from various other sources such as lights, fans, forklifts, labor, etc.
5. A conservative safety factor (10% to 15%) on the total load from 1 to 4.
6. Duration of operation of the refrigeration equipment: generally 18 h/day. The capacity calculated should be preferably increased by the above margin.
7. When the room is used for pre-cooling as well, it is generally preferable to have two or more refrigeration units to meet the total requirement rather than one large unit. This will facilitate using equipment as needed.

Calculation of Building Transmission Load: Refers to Heat Gain / Loss through the Building Walls, Ceiling and Floor

This happens as a result of conduction ($q = kA\Delta T/x$), convection ($q = hA\Delta T$) or combination conduction/convection heat transfer ($q = UA\Delta T$) (refer to section on cooling).

Conduction Heat Transfer through Walls

Example 1: A 10 cm thick brick wall of 10 m² surface area is exposed to warm air outside and cold air inside. The outside wall temperature is 30°C and inside wall temperature is 10°C. What is the steady state rate of heat gain from outside through the wall? The thermal conductivity of the wall is 0.2 W/mC.

$$q = kA\Delta T / x$$
$$= 0.2 \times 10 \times (30 - 10) / 0.1 = 400 \text{W}$$

Example 2: If the inside air temperature is 5°C and the outside air temperature is 40°C, what are the associated convective heat transfer coefficients inside and outside wall surfaces? The rate of heat transfer remains the same.

$$q = hA\Delta T$$

for inside:

$$400 = h \times 10 \times (10 - 5)$$

or

$$h = 400/(10 \times 5) = 8 \text{ W/m}^2\text{C}$$

for outside:
$$400 = h \times 10 \times (40-30)$$
or
$$h = 400/(10 \times 10) = 4 \text{ W/m}^2\text{C}$$

Example 3: Under the situation described in Example 1 and 2 what is the overall heat transfer coefficient?

$$\text{Overall heat transfer coefficient} = U$$

The area of heat transfer surface being same (10 m²),

$$1/U = (1/h)\text{outside} + (x/k)\text{wall} + (1/h)\text{inside}$$
$$(1/U) = (1/4) + (0.1/0.2) + (1/8) = 0.25 + 0.5 + 0.125 = 0.875$$
$$\text{Therefore, } U = 1/0.875 = 1.143 \text{ W/m}^2\text{C}$$

The overall rate of heat transfer $q = UA\Delta T$

where A = area of cross section and ΔT = total temp difference

$$q = 1.143 \times 10 \times (40-5)$$
$$= 400 \text{ W}$$

which is what we calculated from Example 1 and used in Example 2. In this case, we could have determined U also from

$$q = UA\Delta T$$
$$400 = U \times 10 \times (40-5)$$
$$U = 400/(10 \times 35) = 1.143 \text{ W/m}^2\text{C}$$

Example 4: If we add a layer of insulation, 5 cm thick foam, with a conductivity value of 0.015 W/mC, what will be the reduction in the heat transfer rate, assuming that we are still dealing with a 10 m² area and inside and outside temperatures of 5 and 40°C?

The overall heat transfer coefficient U will be different this time.

$$(1/U) = (1/h)\text{outside} + (x/k)\text{wall} + (x/k)\text{insulation} + (1/h)\text{inside}$$
$$= (1/4) + (0.1/0.2) + (0.05/0.015) + (1/8)$$
$$= 0.25 + 0.5 + 3.33 + 0.125 = 4.21$$

Therefore
$$U = 0.238 \text{ W/m}^2\text{C}$$
$$q = UA\Delta T$$
$$= 0.238 \times 10 \times (40-5) = 83.3 \text{ W}$$

Thus the heat gain is reduced to about one fifth of the previous value, a more than 80% reduction as a result of insulation.

In examples dealing with cold storage, generally the overall heat transfer coefficient may be given with an associated surface area and an overall temperature difference from which one can calculate the heat gain or loss.

Calculation of Air Exchange Load Contributed by Opening of the Doors and Ventilation

Every time the door of the cold storage room is opened, a certain amount of heat will be added to the room from outside. The same thing happens when the room air is ventilated, by bringing in fresh air from outside. Usually estimates are provided in the form of equivalent air exchanges based on the room volume and how much heat is gained per unit volume (which depends on the temperature difference between outside and inside).

Example 5: Calculate the heat gain through the opening of doors and ventilation in summer harvest season and fall storage time. During the harvest time, when the produce is being loaded into the room, the air exchange is estimated at 8 complete changes per day and during the subsequent storage it is 3 changes per day. The room volume is 300 m^3 and during the summer time the heat required to lower the outside temperature to room temperature is 80 kJ/m^3 and during fall it is 10 kJ/m^3.

Summer:
 Harvesting & Loading: Heat gain = $8 \times 300 \times 80 = 192,000$ kJ/day
 Storage: Heat gain = $3 \times 300 \times 80 = 72,000$ kJ/day

Fall:
 Harvesting & Loading: Heat gain = $8 \times 300 \times 10 = 24,000$ kJ/day
 Storage: Heat gain = $3 \times 300 \times 10 = 9,000$ kJ/day

Remember:

The heat required to lower the temperature per unit volume (kJ/m^3) also can be calculated if we know the heat capacity of the air (Cp in kJ/kgC), temperature difference (ΔT in °C) and the specific volume (m^3/kg) or density (kg/m^3) of air inside the room [Heat required (kJ/m^3) = ($Cp \times \Delta T \times$ Density) or ($Cp \times \Delta T$/Specific volume)].

Calculation of Product Load

(a) Product load heat is contributed by the product during cooling.

This is the removal of sensible heat from the produce (at the appropriate rate of cooling). It is a major contributor to the cooling load in the pre-cooling operation. It also should be accommodated in the cold-storage if the product is not previously cooled to the storage environment.

As before, the quantity of heat to be removed can be calculated using the equation:

$$Q = mCp\Delta T$$

This includes produce, boxes, crates and any other material that comes into the cold storage with the product. Most cold storage systems are designed to take into account an unexpected load in produce to be brought in; however, as a general rule, objects must be pre-cooled by an appropriate technique before bringing them inside the cold room.

Example 6: A cold storage is designed to operate at 5°C. 10 tons of cabbages at 35°C were moved into the storage room at 200 kg per wooden box, and were expected to be cooled to 5°C in 24 hr. Assume that the weight of each box is 50 kg and heat capacities of separately stored apples and wooden boxes are 4 kJ/kgC and 2 kJ/kgC, respectively. Calculate the amount of heat to be removed in 24 hr (while cooling the product from 35°C to 5°C).

Q (cabbages) $= mCp\Delta T$
$\qquad\qquad\qquad = 10,000 \times 4 \times (35-5) = 1,200,000$ kJ
Q (boxes) $= mCp\Delta T$
$\qquad\qquad\qquad = 50(10,000/200) \times 2 \times (35-5) = 150,000$ kJ

(b) Heat input due to respiration of the produce

Heat from respiration also has been worked out above. Heat of respiration can be obtained by multiplying the respiration rate by an appropriate factor, or can be given in heat units.

Example 7: In the above example, assume that the average temperature of apples undergoing cooling is 20°C during the pre-cooling (24 h). The heat of respiration for apples at 20 and 5°C is given as 5000 and 1000 kJ/ton/day, respectively. Calculate the respiration load from heat during cooling and during subsequent storage.

$$\text{During pre-cooling} = Q = 5000 \times 10 = 50{,}000 \text{ kJ/day}$$

$$\text{During storage} = Q = 1000 \times 10 = 10{,}000 \text{ kJ/day}$$

Remember:

Rather than using one average value of temperature, it can be integrated over shorter time intervals—example, during first 8 h T_{avg} = 25°C, second 8 h, T_{avg} = 15°C and the third 8 h, T_{avg} = 8°C. If we know the heat of respiration at 25°C, 15°C, and 8°C (or one value with a Q10), we can calculate the total heat by summation.

Calculation of Miscellaneous Heat Loads

Miscellaneous heat includes heat input from various other sources such as lights, fans, forklifts, labor, etc. These generally are calculated based on the appropriate data given.

Example:

Lights: Power rating × Daily operating time (24h) =

Fans: Power rating × Daily operating time (24h) =

Forklifts: Numbers × Power rating × Daily operating time (24h) =

Men: Power rating × Daily operating time (24h) =

VENTILATED STORAGE

Ventilated storage is common in temperate zone areas, and also in cold countries, which experience below freezing fall and winter temperatures. Generally, these temperate areas experience a large difference in day and night temperatures. Also, temperatures after fall harvest usually are favorable for such products as potatoes, onions, cabbage and carrots. The products generally are stored in bulk, filling the entire room. Alternately, they may be stored in bins stacked in an orderly fashion. The chamber consists of insulated structures with appropriate designs for ventilation. When the temperatures are favorable, the units are ventilated to replace the warm inside air with cool fresh air. In winter climates when the outside temperatures are much cooler than tolerated by the produce, the air in the cold storage is mixed only partially with the outside fresh air, just enough to bring down the temperature to the required levels. Modern ventilated storage makes use of automatic control of outside air intake, mixing it with the inside air and re-circulating it after necessary humidification. Recirculation is essential to conserve the humidity of the system. Air circulation throughout the product also is essential to prevent localized heating of the product due to build-up of heat of respiration, which can lead to spoilage. Special distribution channels must be provided along the floor with adequate vent-holes for forced circulation of air through the stack. These can be simple inverted

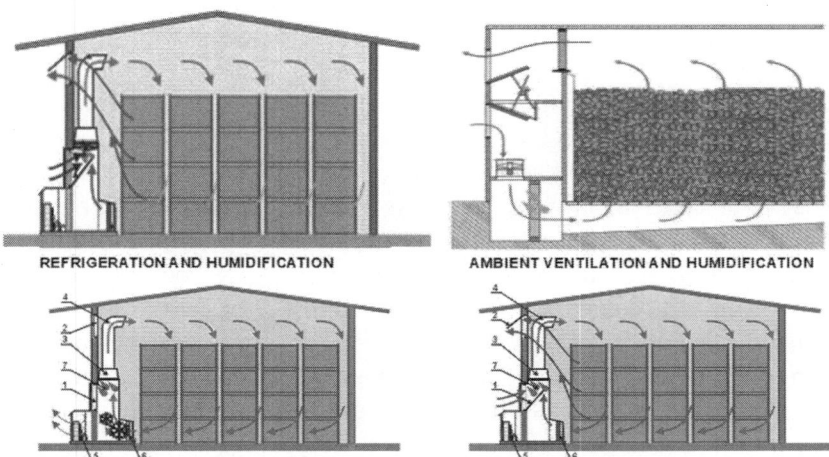

FIGURE 10.2. Ventilated and refrigerated bulk storage systems (Courtesy: Agroel, s.r.o. Dobrovice, CZ).

large diameter half pipes with holes drilled at appropriate locations to divert the air flow. In summer months, there generally will be additional mechanical refrigeration used on a standby basis to meet needs. In general, the material from these units will begin to be removed for packing and distribution around that time. Often, there will be flume channels along the floor to move the produce through a water flume.

An example of a well-structured ventilated storage facility is illustrated in the Figure 10.2. One of them is for bulk storage of onions with ventilation and humidification. Others illustrate modified systems for storage of products like onions, potatoes and carrots in open bulk containers rather than directly filling the rooms. Systems also are illustrated with additional refrigeration and humidification which can be used as regular atmosphere refrigerated storage commonly employed for most produce.

REFRIGERATED (REGULAR ATMOSPHERE, RA) STORAGE

Regular atmosphere refrigeration units are insulated vapor proof structures with mechanical refrigeration units. These are suitable for any fruit or vegetable, provided their optimal storage conditions are maintained. An appropriate thermostatic control is used to regulate the temperature. Refrigeration capacity is appropriately designed to serve as continuous storage of a variety of fruits and vegetables. Generally, some provisions are made in the refrigeration design to take a certain amount for fresh product directly from the field (to accomplish cooling as well). However, as a general rule, the produce is pre-cooled prior to filling the cold storage facility. A refrigerated system for bulk bin storage is also shown in Figure 10.2.

Some advantages of a conventional refrigeration system include: easy to install, maintain and run; easy to find qualified service personnel; gives good control of air temperature; allows prompt cooling in a warm harvest season, and extends the marketing season (http://www.omafra.gov.on.ca/english/engineer/facts/98-073.htm). On the other hand, disadvantages of a conventional refrigeration system include: expensive to install, maintain and run (relative to ventilated storage); can dry out the air and shrink the produce; difficult to add humidity to the air and can freeze the produce in top bins. A typical commercial storage room is illustrated in Figure 10.3.

In order to provide a high relative humidity environment, several techniques and modifications are used: jacketed storage, Filacell® system or fog jet humidification.

Jacketed Storage

Jacketed storage is a modified storage procedure for producing high humidity in the storage environment. It builds up humidity by conserving the moisture lost by the produce due to transpiration. It consists of a sealed chamber (jacket) to hold the produce. The jacket is externally cooled by air coming from the refrigeration system. This air does not contact the produce (indirect cooling). The produce will undergo transpiration as usual initially and lose moisture to the surrounding air. Since the room is sealed, the moisture is retained in the air and does not get carried away outside. This results in an increase in the relative humidity, which eventually reaches saturation levels and stops the transpiration. Since it is a closed system, care must be taken to admit the necessary amount of humidified air into the system to maintain aerobic respiration conditions. Secondly, accumulated CO_2 also needs to be removed either by ventilation or using a scrubbing source such as lime. Placing a bag of lime in the chamber will do the job. Cooling inside the jacket is slow and hence during the initial stages the jacket can be kept open to let the cold air in and adequately cool the product. After that not much refrigeration is required since the jacketed structure is within the cold room except for the heat produced by respiration. A typical jacketed system is shown in Figure 10.4.

FIGURE 10.3. Refrigerated storage for kiwi fruits—inset kiwi in retail boxes and bulk bins (Source: http://evolutionkiwi.com/id2.html).

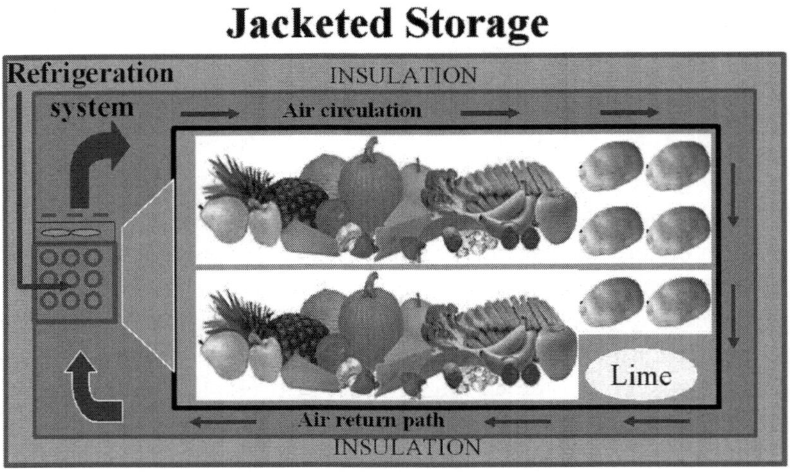

Jacketed Storage

FIGURE 10.4. Schematic representation of a jacketed storage system.

Filacell® System

Filacell® is another modification in refrigerated storage systems to produce high humidity in the chamber and reduce transpiration losses. It is based on the principle of evaporative cooling discussed earlier. The lower the relative humidity of the air, the higher is the transpiration loss. Hence, the transpiration rate can be reduced significantly in high relative humidity storages. The loss of moisture also depends on temperature fluctuations in the chamber. But maintaining high humidity will prevent the air from carrying the moisture to the cooling coils. The Filacell system is based on humidifying the air that already is cooled by the refrigeration system, thus entering the chamber fully saturated (Figure 10.5). The Filacell unit is similar to the evaporative cooling set up discussed in the previous chapter. It consists of an expanded web like structure, which is kept wet by a water spray, and the cold air leaving the coils is forced up through the unit. Hence, the air is saturated with water vapor before entering the cold storage chamber. However, if the temperature of the cold air is not close enough to the load temperature, the air will undergo warming and will have room to pick up moisture. So it is important to have an appropriate refrigeration capacity to have few temperature fluctuations in the room and to keep the product precooled prior to loading. A simplified version of such concepts can be adapted to any refrigeration system as long as air can be forced through wet porous humidification pads.

Filacell storage

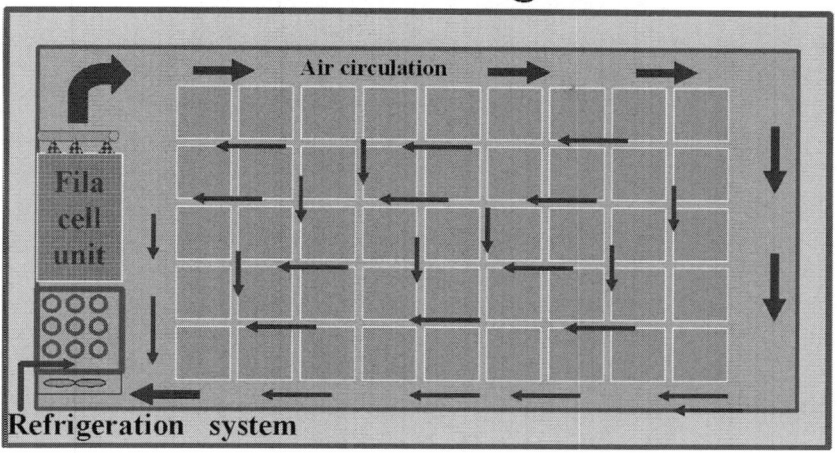

FIGURE 10.5. Schematic representation of a Filacell storage system.

Fog Jet Humidification

Several commerical tools and devices can be used to create a fog or fog jet. See Figure 10.6. These are simply high-pressure spray devices that dispense water in the form of fine droplets through specially con-

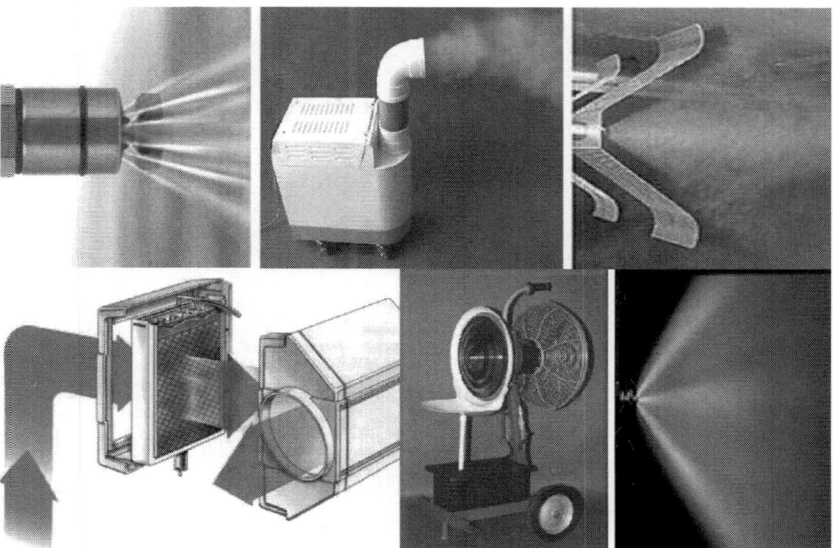

FIGURE 10.6. Simple techniques for crating fog /mist for humidifying air.

structed nozzles. Depending upon the intended purpose, the nozzle hole is varied. When intended for a high-power water jet, the hole would be a bit larger, and to make a finer mist microholes are used. Often multiple jets are used to create a dense fog. The fog serves to add humidity to the room. The fine mist of water particles instantaneously vaporize creating an evaporative cooling scenario, thereby simultaneously cooling and humidifying the air. Generally, the fog is sprayed onto an air stream in front of a fan so that it can be efficiently mixed. To spray water into evaporating pads or filacell systems, a coarser spray can be employed. For most domestic and office climatisation purposes, an ultrafine mist can be created by combining the system with other devices such as ultrasonics.

Humidifying Pads

Commercial set ups are available for humidification systems that work without producing water droplets. Water evaporates from special cross-fluted porous pads that are continuously saturated with water while air passes through their unique structure. Only enough water is supplied to the air as is able to evaporate. No water droplets are supplied to the storage area and therefore the risk of microorganisms spreading (e.g., molds) on wet produce is significantly reduced. The saturation

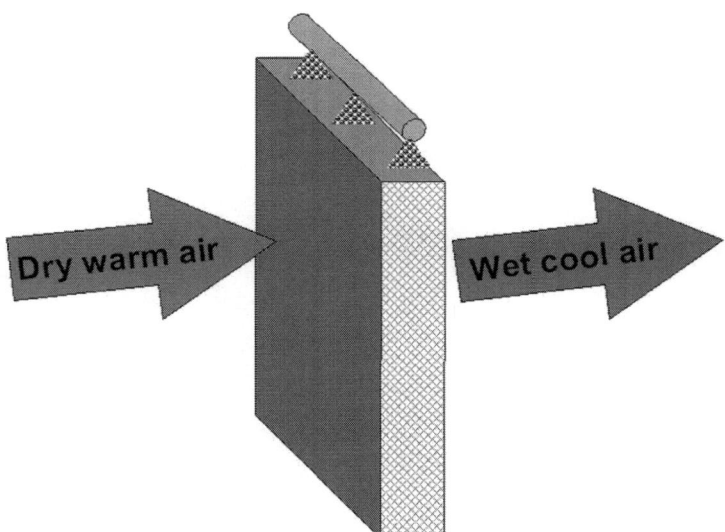

FIGURE 10.7. A simple humidification pad.

efficiency of humidifying pads is very high. The use of humidifying pads provides more benefits as well. Evaporation of water also results in cooling the air that passes through the pads. This simple physical principle leads to energy savings. The cooling effect feature also is useful in ventilated storages (Figure 10.7).

CONTROLLED ATMOSPHERE (CA) STORAGE SYSTEMS

Controlled atmosphere storage has been referred to as one of the most important innovations in fruit and vegetable storage since the introduction of mechanical refrigeration. In conjunction with low temperature and high relative humidity, this method involves the alteration of the gaseous environment inside the storage chamber. Oxygen concentration generally is lowered to about 2–3%, and carbon dioxide is added up to 5%. As discussed in the chapter on respiration, both of these help to retard the rate of respiration, the former by limiting one of the reactants (oxygen), and the latter by accumulating a product of respiration (CO_2). In addition, lowering of temperature obviously has a significant effect on lowering the respiration rate. These three factors taken together have a synergistic effect, thereby providing opportunity for enhancing the postharvest shelf life by a margin greater than that possible by the added effects of the individual components. Sometimes, CA storage also may entail the addition of other gases, such as carbon monoxide, to provide additional fungistatic effects. Ethylene, a growth promoter, and a natural product of respiration, is also removed.

Several advantages have been recognized for products in CA storage:

1. Lowered respiration rate.
2. Retarded senescence.
3. Suppressed ethylene production rate.
4. Reduced fruit sensitivity to ethylene.
5. Improved retention of green color.
6. Improved texture.
7. Improved retention of nutrients like ascorbic acid.
8. Alleviation of physiological disorders like russet spots on lettuce, internal breakdown.
9. Suppressed activity of pathogens.

10. Control of insect activity (only possible at high CO_2 concentrations, not usually tolerated by fruits and vegetables).

Some undesirable effects observed are:

1. Elevation of physiological disorders in certain produce, such as black heart of potatoes or brown heat in some varieties of apples and pears.
2. Irregular ripening: banana, pear.
3. Development of off flavors, especially at low levels of oxygen.
4. Stimulation of sprouting and retardation of periderm development in some tubers like potatoes.

As CA storage takes place in sealed rooms for extended periods of time, there is the possibility for enhancing production of metabolic and other volatiles. These arise not only from the fruits, but also from degradation of wooden bins, the growth of molds, etc. In normal ventilated rooms, the concentration of such volatiles will be low. Generally, these volatile accumulations will be lower in CA systems equipped with activated charcoal scrubbers intended for removal of ethylene and excess carbon dioxide. Under low oxygen tension as in CA rooms, acetaldehyde and ethanol production can result from partial anaerobic fermentation. Such volatiles can be removed from the fruits after they are taken out of the CA room and stored under regular atmosphere for some time.

It should be noted that the CA storage is really intended for long-term storage and unlike the regular atmosphere storage the produce is not generally taken in and out at frequent intervals. Further, the environment inside the storage does not support human life, and therefore the people who enter CA storage area and work there need to wear an oxygen mask to enter.

Again, CA storage is not practical for all fruits and vegetables. It is expensive to construct and operate a CA room and requires specialized structures, equipment and control systems. Highly skilled workers are required for both operation and maintenance. It would add significantly to the product cost and therefore it is used for only those commodities for which the additional cost is acceptable.

Table 10.2 indicates the fruits and vegetables for which the use of CA storage offers potential benefits (Kader, 1985). Unless the benefits are significant, the added cost for the storage may not be justified.

TABLE 10.2. Commodities that Demonstrate a Potential for CA Storage with Some Commercial Success.

Fruits	Apple, kiwi, pear, strawberry, dried fruit, banana, cherry, fig, nectarine, peach, plum, prune, avocado, lime
Vegetables	Asparagus, artichoke, broccoli, cabbage, cauliflower, corn, leeks, lettuce, onion, tomato

CA Generation and Control

There are two issues concerning CA systems: (1) CA generation and (2) CA control. The generation implies lowering of oxygen concentration and elevating carbon dioxide to levels appropriate for the CA room operation (generally 2–3% O_2 and 5% CO_2). This can be done by two methods: (1) passive process and (2) active process. It is well recognized that for produce respiration, oxygen is required and CO_2 is released as a product of respiration. Hence, in a closed CA room, if the produce loaded into the room is allowed to respire, oxygen available in the room (air has about 21% O_2) will be gradually used up and CO_2 will begin to accumulate. With ongoing produce respiration, oxygen concentration in the room is continuously lowered and carbon dioxide level is continuously increased. By the time 5% CO_2 is accumulated, oxygen concentration will decrease by about 5% (assuming an RQ of 1.0). While O_2 needs to be lowered further, CO_2 has reached the limit. From now on the CO_2 needs to be maintained at this level by removing excess CO_2. Further respiration would eventually deplete sufficient O_2 to reach the target level and then the amount of oxygen required for respiration needs to be added to the room to maintain the 2% concentration. This process is called passive modification of the CA environment and is a very slow process. In the active process, the storage atmosphere is modified using rapid means to lower the oxygen level through oxygen control systems and quickly elevate the carbon dioxide level by adding it through a gas cylinder. Then they need to be controlled. For oxygen level control, the required amount of O_2 is maintained by adding a calculated amount of air and for maintaining the CO_2 level, a CO_2 control system is periodically activated to remove CO_2 from the room. Alternately a membrane system can be used for both creating and maintaining the CA, which will be discussed later. Also, a hypobaric system to create and maintain CA, is discussed later.

Lowering of Oxygen

External Gas Generators

External gas generators operate either on an open-flame or catalytic burner to remove O_2 from the incoming air. The heat added by the burners needs to be removed, which is done by a water spray that cools the air and humidifies at the same time. The disadvantage of these types of system is that fuel is required and CO_2 is produced as a product of combustion and must be removed. Filtration may also be necessary before letting air back into the storage room. The advantages of these systems are flexibility of operation and the rapidity with which O_2 is depleted in the initial stages of storage. The catalytic system is preferable in ensuring better combustion and fewer combustion by-products. Catalytic burners are more expensive to install. The operation can be carried out in either a re-circulating configuration or purge system to flush out the oxygen. In the re-circulating type, the CA air is pulled out, sent through the burners and readmitted after depleting the O_2. This will continuously deplete the oxygen from the room until the desired level is reached. In the purge system, fresh air from outside is fed to the gas generators, and the oxygen-free air after cleaning and humidification is continuously fed into the CA room and vented out from the opposite end of the room. As continuously oxygen-free air is entering and oxygen rich air is leaving, the oxygen concentration will progressively decrease until reaching the target. The re-circulating and purge-type flows are illustrated in Figure 10.8.

Nitrogen Flush

Flushing the storage room with N_2 is an excellent method for depleting O_2 in CA. This can be done using pressurized N_2 gas cylinders. Alternately, liquid nitrogen (LN) also can be used by spraying the LN in front of evaporator blowers. Care must be taken to make sure too

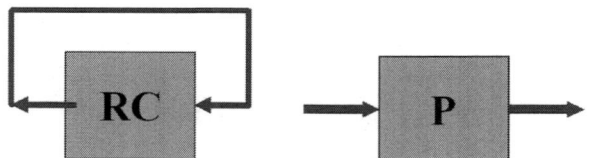

FIGURE 10.8. Schematics of re-circulating and purge systems.

much LN does not enter the CA room as it can compromise temperature maintenance. An O_2 sensor is used to determine when to stop the LN addition.

Elevating CO_2 level

If not done by produce respiration, CO_2 is increased by introducing a calculated volume of CO_2 to the CA room from CO_2 gas cylinders.

Maintaining CA

In order to maintain controlled atmosphere conditions, the exact amount of O_2 required for respiration needs to be admitted into the CA room. This is done by calculating the oxygen consumption rate by the produce and the introduction of oxygen (air) into the room through a metering device. Alternately, an inlet device can be activated once the oxygen level reaches the low level target and until air admitted reaches the upper level set point. The on-off will be operated between the two limits. For CO_2, the excess CO_2 produced need to be removed. This is done likewise by activating a CO_2 scrubbing device once the upper limit of CO_2 is reached and scrubbed until it reaches the lower set point and stopped. Again, the on-off will be operated between these two limits. The following are available scrubbers for CO_2.

Scrubbing of CO_2

Several methods have been used to scrub CO_2: caustic soda, hydrated lime, water, activated charcoal and molecular sieves. Monitoring of the atmospheric composition is necessary to control the scrubbing system.

Caustic soda (NaOH) was one of the first reagents used commercially for CA storage applications and was used in mixture with water. The solution was circulated in open tubes and thus absorbed CO_2, the residence time being controlled according to the required removal rate. Its use was discontinued due to its corrosiveness and potential danger in handling. However, dry caustic material has been proposed as a reasonable alternative.

Perhaps the simplest method of controlling CO_2 levels is through hydrated lime scrubbers ($Ca(OH)_2$). The scrubber is an insulated and sealed plywood box usually containing between half to the total amount of lime required to remove the CO_2 produced during the entire storage period. The box is connected to the CA room and airflow to it is left to natural convection or controlled by forced-air blowers and dampers.

There are two types of water scrubbers for CO_2. One is the brine scrubber in which brine is pumped into an aerator, which causes CO_2 to escape to the atmosphere. Dry evaporators' coils are used to prevent corrosion. A modified system also is available using two aerators, one internal and one external.

Activated charcoal and molecular sieve scrubbers are units filled with the respective absorbers, two blowers and four timer-controlled valves. Air from the CA room is circulated to the scrubber where CO_2 is absorbed. The CO_2-depleted air is returned to the storage room. When the absorbent is saturated, outside air is circulated through the scrubber and back to the outside to remove the CO_2. If a molecular sieve is used, it is also heated to speed up the reactivation process.

ETHYLENE IN POST-HARVEST TECHNOLOGY

Ethylene is the simplest alkene with a chemical formula C_2H_4. It is biologically very active even in trace amounts and acts physiologically as a hormone in plant products like fruits and vegetables. It acts at trace levels throughout the life of the plant by stimulating or regulating the ripening of fruit, the opening of flowers, and the abscission (or shedding) of leaves. It also induces several physiological disorders and is desirable in ripening, but undesirable in storage. Ethylene gas (C_2H_4) is a nearly odorless (slightly sweet smell of ether), colorless gas that exists in nature. Solubility in water is 20 mg/L (20°C). Its threshold level is 0.1–1 ppm. Excessive inhalation results in loss of consciousness. It is flammable in the 3–32% range; however, the usage level is very low: 10 to 100 ppm (0.001–0.01%). Acetylene has some ethylene mimicking power.

Historically, ethylene activity has been utilized for centuries. Many fruits like sorb apples, burnt straw, etc. have been recognized to give off active gas, which can trigger ripening of fruits. Kerosene stove gas traditionally has been used for ripening. So has dry straw. Mango in Asian countries is traditionally ripened on a bed of straw, which produces ethylene, and the storage of fruits in a closed room adds further ethylene to enhance the ripening. In the 1920s, the stove gas was identified as ethylene and since 1930s commercial use of ethylene has been practiced. Today, ethylene is FDA approved as a commercial ripening agent, and no-residue standards are in place indicating it to be safe. The ripening, however, needs to be carried out under carefully controlled conditions with trained operators.

Ethylene Ripening: Optimum Ripening Conditions for ethylene ripening are: 18–25°C, 90–95% RH, ethylene concentration 10–100 ppm and treatment time, 24–72h. During the treatment, good ventilation is necessary to prevent CO_2 build-up, because excess CO_2 inhibits both ethylene synthesis and activity. Since ethylene can be flammable at higher concentrations, only water or steam-heating coils are used for temperature control rather than electrical ones. All sources of flames or sparks are eliminated.

For ripening purposes, ethylene can be introduced into the chamber through ethylene gas cylinders (measured amounts are metered into the room). A commercial ethylene mixture or ripe gas, which is a mixture of 6% ethylene in CO_2 also can be used. This is called explosion proof because of the presence of CO_2. Ethylene also can be commercially produced from ethyl alcohol or other products of ethylene like Ethepon. Many commercial products are available as sources of ethylene, for example: Easy Ripe® Ethylene Generator, Ethy-Gen® II Ripening Concentrate (Catalytic Generators, LLC. Norfolk, VA).

Ethylene in Postharvest Storage: While ethylene gas is used under controlled conditions as a ripening agent, even small amounts of ethylene gas during shipping and storage cause most fresh produce to deteriorate faster. A single propane-powered forklift can cause serious damage in highly ethylene gas-sensitive commodities. Undesirable consequences of ethylene use are: accelerated senescence, accelerated ripening, loss of green color in vegetables, acceleration of physiological disorders, such as russet spotting on lettuce, bitterness in carrots, sprouting of potatoes and core browning of apples. Hence, it is highly essential that ethylene be removed from postharvest storages.

The biochemical pathway leading to the synthesis of ethylene is indicated below:

$$\text{Methionine} \rightarrow \text{s-adenosylmethionine (SAM)} \rightarrow$$
$$\text{1-aminocyclopropane-1-carboxylic acid (ACC)} \rightarrow \text{ethylene}$$

Chemicals that have been used to inhibit the synthesis of ethylene (block synthesis SAM and ACC) are: aminoethoxyvinyl glycine (AVG), methoxyvinyl glycine (MVG) and aminoacetic acid (AOA). Others that have been used to inhibit the action of ethylene are: silver thiosulfate (STS), carbon dioxide (CO_2) nickel (Ni), and 1-methylcyclopropane

(MCP). MCP is a gas that can saturate ethylene receptor sites and block action for several days.

Strategies for removing ethylene from the storage room include: (1) eliminate all sources of ethylene, such as ripening fruits, smoke, internal combustion engines (gas forklifts), decomposing produce, rubber materials exposed to heat, etc.; (2) provide good ventilation, which can remove excess CO_2 as well as ethylene; (3) use a chemical scrubber. Scrubbers for ethylene include: potassium permanganate ($KMnO_4$), which oxidizes C_2H_4 to CO_2 and H_2O; activated charcoal filters which can absorb ethylene; ozone—ozone oxidizers C_2H_4 to CO_2 and H_2O. However, direct exposure to ozone may be harmful to produce. The air from the storage room can be treated with a calculated amount of ozone and the treated air readmitted to the room. A palladium (Pd) promoted powdered material has been identified to have a far greater ethylene adsorption capacity than KMnO4-based scavengers when used in low amounts and in conditions of high relative humidity (RH).

Ethylene can also be removed using ethylene-absorbing filters. These have been proven in reducing and maintaining low ethylene levels. If ethylene damage is suspected, a quick and easy way to detect ethylene levels is with handheld sensor tubes. This will indicate if the above steps should be followed. Ethylene-absorbing sachets are commercially available. These can be placed in storage units or in bulk containers intended for long-range transportation. Ethylene producing fruits (such as apples, avocados, bananas, melons, peaches, pears and tomatoes) should be stored separately from ethylene-sensitive ones (broccoli, cabbage, cauliflower, leafy greens, lettuce, etc.). 1-Methylcyclopropene (MCP), an inhibitor of ethylene action, has been recognized to delay the ripening and senescence of many climacteric fruits, including apple. As such, MCP reduces respiration and ethylene production, and slows softening and deterioration of apple fruit. MCP inhibits the action of ethylene by blocking ethylene receptor sites in the plant cell.

MEMBRANE SYSTEM FOR CA/MA

Controlled and modified atmosphere (CA, MA) storage units are used in the same context and usually they mean the same in the context of a storage room. They both involve changing the storage atmospheric composition to a desired level. The controlled atmosphere refers to an atmosphere with a strict control of the gas concentrations of oxygen, carbon dioxide and nitrogen. Modified atmosphere (MA) is used more in situations where the produce is held under conditions where the atmosphere is modified by the produce, package, over wrap, box liner or pallet. Oxygen generally is reduced through respiration by the produce (passive system), and the carbon dioxide level is determined by the permeability of the film, respiration, temperature, gas barrier property and other factors. The concept of modified atmosphere was developed almost simultaneously as a packaging concept and extended to larger containers and small storage units. In MA designated systems, the carbon dioxide and oxygen levels are not strictly controlled to specific concentrations.

The membrane system is a type of modified storage that has evolved as an extension of a modified atmosphere package (MAP). The MAP's commercial success and CA's scientific background provided the basis for the membrane system for CA generation and maintenance. It is based on selective diffusion of gases through the membrane material, which at a predetermined product loading will bring in enough oxygen for the aerobic respiratory need and at the same time help to remove the carbon dioxide produced by the respiratory process. It establishes a dynamic equilibrium between produce respiration and membrane permeation. The theory is that when the right film and right temperature are used, the membrane will help to maintain a theoretically correct and beneficial mixture of carbon dioxide and oxygen inside the chamber. The method was invented by two French investigators Marcellin and Leteinturier, and involves the use of elastomers of silicone (Smock, 1979).

The membranes breathe in oxygen and breathe out carbon dioxide in synchrony with the produce respiration pattern and hence maintain a preset concentration of oxygen and carbon dioxide inside the room. Since most CA/MA systems operate at considerably reduced oxygen levels (e.g., 2%) and slightly elevated CO_2 levels (e.g., 5%), the concentration gradient for these two gases across the membrane are very different (19% for O_2 and 5% for CO_2 in the above assumption). Hence, the membrane has to have a selective differential permeability to these gases in order to maintain a set atmospheric composition in the storage room. The simplified concept is detailed below, and the complexity of the concept also is elaborated.

A membrane system is based on the gas diffusion process. It uses special membranes with selective permeability. The gas diffusion occurs between the air from inside and outside the chamber across the semi-permeable special membrane. The extent of gas diffusion depends on the concentration gradient of the components (primarily O_2 and

CO_2) and permeability of the membrane to these gases. The resulting concentration depends on: (1) permeability of the gas through the membrane, (2) concentration gradients, (3) respiratory demand of the produce (respiration rate and quantity of the produce), and (4) temperature.

Example: Consider a semi-permeable membrane as a window on a CA room. The concentration gradients across the membrane (inside and outside the CA chamber) for a CA room maintained at 2% O_2 and 5% CO_2 can be represented as shown in Figure 10.9.

Fick's Law states that the gas flux (J) across the membrane is proportional to the concentration gradient (ΔC) and permeability (P):

$$J \propto \Delta C \times P = \text{Constant} \times \Delta C \times P = K \cdot C \cdot P$$

The concentration gradients according to the previous example are:

O_2 $\Delta C = 21 - 2 = 19\,\%$ going inward

CO_2 $\Delta C = 5 - 0 = 5\,\%$ going outward

Different Scenarios for Membrane

Membrane with Equally Permeability to both O_2 and CO_2

Since the membrane is equally permeable, the flux (J) rate for O_2 and

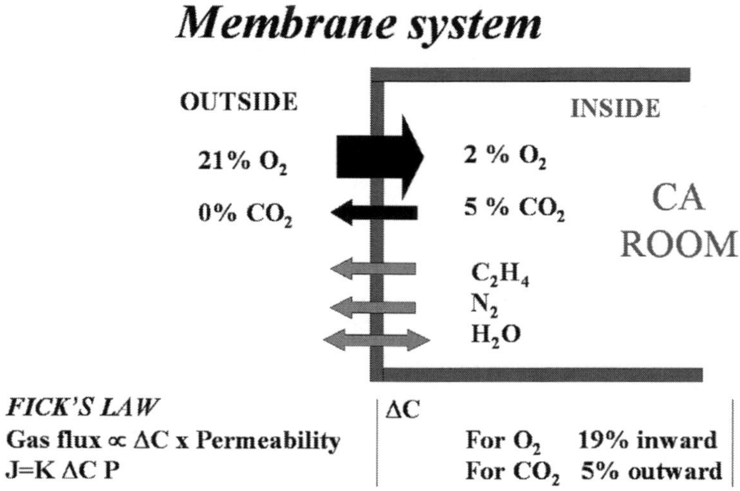

Membrane system

OUTSIDE		INSIDE
21% O_2	→	2 % O_2
0% CO_2	←	5 % CO_2 CA ROOM
		C_2H_4
		N_2
		H_2O

FICK'S LAW	ΔC	
Gas flux $\propto \Delta C$ x Permeability		For O_2 19% inward
J=K ΔC P		For CO_2 5% outward

FIGURE 10.9. Concentration gradient of gases across a membrane in a membrane CA system.

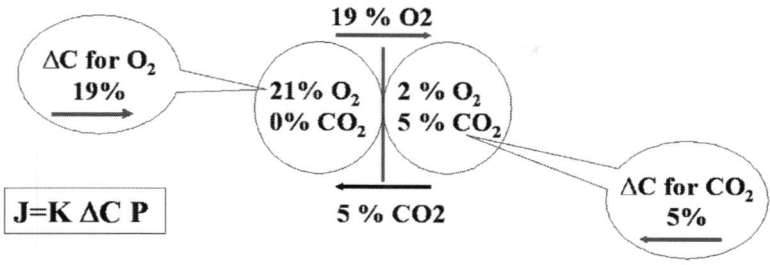

- Oxygen entry rate > carbon dioxide exit rate
- Based on ΔC, O_2 Inward Flow Rate = CO_2 Outward Flow Rate x (19/5)

FIGURE 10.10. CA system with a membrane equally permeable of gases.

CO_2 depends primarily on their concentration gradients. Relatively, the rate at which O_2 enters the room will be higher than the rate at which CO_2 will exit from the room. The proportion of the two rates will be $O_2:CO_2 = 19:5$.

When there is a certain quantity of respiring produce inside the chamber, it will absorb O_2 and release CO_2 via respiration. Let us assume a simple case of carbohydrate substrate with an RQ of 1.0.

i.e., O_2 absorbed = CO_2 released

Let us further assume that we have used an appropriate size of membrane for a predetermined quantity of produce so that the amount of O_2 coming in through the membrane is exactly the same as the amount of O_2 being absorbed by the produce for respiration.

Question: In this scenario, what will be the resulting equilibrium CA conditions inside the storage chamber (what are the O_2 and CO_2 concentrations)?

Answer: Since the amount of oxygen coming in is exactly matched with the oxygen being absorbed by the produce, there will not be any change in the oxygen concentration and hence it will be maintained at a 2% level. However, the same is not true with the CO_2 concentration. CO_2 is produced at the same level as O_2 consumption; however, the CO_2

exit rate is far slower than the oxygen entry rate (see the ratio indicated above). This will lead to incomplete removal of the CO_2 produced by the respiratory process in the room, resulting in its accumulation. This will continuously increase the concentration of CO_2 in the room, which continuously increases the CO_2 exit rate (because of an increase in ΔC for CO_2) until the ΔC for CO_2 matches with ΔC for O_2 (which is 19% in the above example). Hence, the theoretical equilibrium concentration of CO_2 will be 19% if the produce can withstand that level. Most likely it will result in aerobic respiration and CO_2 damage to the produce, a highly undesirable situation.

Alternately, let us assume that amount of CO_2 coming out of the room through the membrane matches the amount of CO_2 produced by respiration.

Question: In this scenario, what will be the resulting equilibrium CA conditions inside the storage room (what are the O_2 and CO_2 concentrations?)

Answer: Since the amount of CO_2 coming out of the storage room is exactly matched with the CO_2 produced by the respiring produce, there will not be any change in the CO_2 concentration and hence it will be maintained at the 5% level. However, the same will not be true with the O_2 concentration. O_2 is consumed at the same level as CO_2 production; however, the O_2 entry rate across the membrane is far greater than the CO_2 exit rate (see the ratio indicated before). This will lead to excessive entry of O_2 into the chamber than required by the respiratory process, resulting in its accumulation. This will continuously increase the concentration of O_2 in the room, which continuously decreases the O_2 entry rate (because of a decrease in ΔC for O_2) until the ΔC for O_2 matches the ΔC for CO_2 (which is 5% in the above example). Hence, the theoretical equilibrium concentration of O_2 will be 16%, resulting in a ΔC level of 5%. This oxygen concentration is only slightly below that in the atmosphere and hence it provides a less effective CA from an O_2 point of view, while there will be beneficial effects from the 5% CO_2. Overall, this at least is a better alternative than the earlier scenario because it will not be damaging to the produce.

Membrane with Differential Permeability

Obviously, in the previous example, the difficulty is achieving the desired CA given the differences in the rates of the two active components resulting from their differences in concentrations. A solution is to have a membrane that can compensate for this difference.

Let us assume that the membrane has:

CO_2 permeability [19/5] times the permeability of O_2

Our objective is to have:

Rate of O_2 entry = rate of CO_2 removal

From Fick's law:

$$J = K \Delta C \cdot P$$

Oxygen entry rate = $J_{(O_2)} = K \cdot \Delta C_{(O_2)} P_{(O_2)}$

Now we have

$$P_{(CO_2)} = [19/5] \, P_{(O_2)} \quad \text{or} \quad P_{(O_2)} = [5/19] \, P_{(O_2)}$$

Also

$$\Delta C_{(O_2)} = [19/5] \Delta C_{(CO_2)}$$

Substitute this is in the Fick's law equation:

$$\begin{aligned}
\text{Oxygen entry rate} = J_{(O_2)} &= K \cdot \Delta C_{(O_2)} P_{(O_2)} \\
&= K \cdot [5/19 \, \Delta C_{(CO_2)}] \, [\, 19/5 \, P_{(CO_2)}] \\
&= K \cdot \Delta C_{(CO_2)} \, P_{(CO_2)} \\
&= \text{Carbon dioxide removal rate}
\end{aligned}$$

Now what will be the resulting CA conditions inside?

You now can match the oxygen consumption rate or the carbon dioxide release rate. Either way the resulting CA will be 2% O_2 and 5% CO_2. This is because we have the same rate for the oxygen entry and carbon dioxide removal both matching with counterparts through respiration.

Therefore in Membrane CA:

O_2 and CO_2 Exchange through Membrane	=	O_2 and CO_2 Exchange through Respiration

The type of membrane determines the permeability characteristics of the membrane. The size (area) of the membrane and its concentration gradients determine the exchange of gases through the membrane. The type of produce determines its respiratory characteristics. Together with quantity, this determines the exchange of gases through respiration. The CA conditions will influence the gas exchange from both respiration and permeation because they determine (1) the concentration gradients and (2) the respiration rate. Temperature will influence both, because the respiration rate and permeability characteristics are both affected by temperature. The membrane area usually is adjusted to match the exchange of gases through respiration and permeation. Silicone elastomer is a special membrane that has a CO_2 to O_2 permeability ratio of about 5.5: 1. This has been used in many industrial applications of CA.

A simple example of membrane storage follows.

This is an example of a membrane storage for apples (see Figure 10.11). The desired conditions are 2% O_2 and 5% CO_2 at 2°C. Apples in crates are loaded into the room to a full capacity of 10 tons (10,000 kg). Respiration rate of apples at 2°C is given as 3.5 mL CO_2 /kg h and it is assumed that the respiration rate in the CA room is 70% of the normal rate. A window of silicone membrane is to be installed. The window

Design of a membrane storage

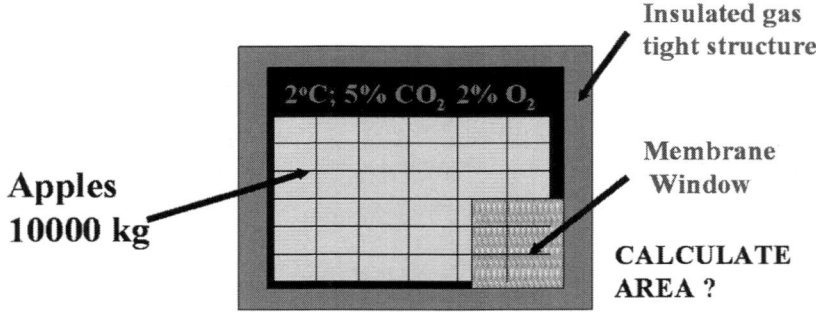

Data:
Respiration rate: 3.5 ml CO_2/kg h (normal); 70% of normal in CA
Silicone membrane permeability (2°C); 1750 L CO_2/m² day atm
350 L O_2/m² day atm

FIGURE 10.11. Design of a CA storage with a membrane window.

material has a permeability of 1750 L CO_2/day atm and 350 L O_2/day atm. Determine the area required based on (1) oxygen consumption rate and (2) carbon dioxide liberation rate. If the two are not same, which one is preferred? What other alternative can be used to attain the desirable conditions?

Figure 10.11 represents data presented in the problem.

1. Calculation based on CO_2 production rate

    First step is to find out the CO_2 production rate based on the data given on respiration.

    According to data: Respiration rate = 3.5 mL CO_2/kg h at 2°C under normal conditions and 70% of this value in the CA room

    CO_2 Produced = $R \times$ Mass
    $\qquad$ = 3.5 mL CO_2/kg h $\times$ 0.7 $\times$ 10000 kg $\times$ 24 h/day
    $\qquad$ = 588,000 ml/day   or   588 L/day

    Next step is to calculate the CO_2 that escapes through the membrane

    This comes from Fick's law

    $$J(CO_2)/\text{day} = P(CO_2)\,(L/\text{day atm m}^2) \times C(\text{atm})$$
    $$= (1750) \times (0.05) = 87.5 \text{ L/m}^2$$

    In order to maintain the CA conditions:

    (CO_2) produced by respiration = (CO_2) expelled by permeation

    Equating the permeation to the respiration

    $\qquad$ 588 L/day = 87.5 L/day m^2

    Therefore area required

    $\qquad$ A = 588/87.5 = 6.72 m^2

    This area can be realized by any combination of length and width that gives the product equal to 6.72 m^2. Such a window would remove from the CA room the exact volume of CO_2 that is produced by respiration. Hence the CO_2 concentration of 5% is maintained.

2. Calculation based on O_2 production rate

First step is to find the CO_2 production rate based on the data given on respiration.

Again according to data: Respiration rate = 3.5 mL CO_2/kg h at 2°C under normal conditions and 70% of this value in the CA room.

Assuming RQ = 1, the oxygen consumption rate would be same as the carbon dioxide production rate.

$$O_2 \text{ Produced} = R \times \text{Mass}$$
$$= 3.5 \text{ mL } O_2/\text{kg h} \times 0.7 \times 10000 \text{ kg} \times 24 \text{ h/day}$$
$$= 588{,}000 \text{ ml/day} \quad \text{or} \quad 588 \text{ L/day}$$

Next step is to calculate the O_2 that comes in to the room through the membrane.

This comes from Fick's law

$$J(O_2)\text{day} = P(O_2) \, (L/\text{day atm m}^2) \times \Delta C(\text{atm})$$
$$= (350) \times (0.19) = 66.5 \text{ L/m}^2$$

In order to maintain the CA conditions:

(O_2) consumed by respiration = (O_2) inlet by permeation

Equating the permeation to the respiration

$$588 \text{ L/day} = 66.5 \text{ L/day m}^2$$

Therefore area required

$$A = 588/66.5 = 8.84 \text{ m}^2$$

Again this area can be realized by any combination of length and width that gives the product equal to 8.84 m². Such a window would add to the CA room an exact volume of O_2 that is consumed by respiration. Hence the O_2 concentration of 2% will be maintained.

Area required is:

6.72 m² for maintaining CO_2 concentration in the room at 5%
8.84 m² for maintaining O_2 concentration in the room at 2%

You can't have both because there is only one window and it can have one area.

The one that is appropriate for O_2 is not appropriate for CO_2 and vice versa. Which area would you use? 6.72 m² or 6.72 m²?
What are the consequences of your selection?
What is the problem?
What alternative can you think of to provide a good solution?

HYPOBARIC SYSTEM FOR CA/MA

Hypobaric is another form of CA/MA and has been given several names: hypobaric storage, low-pressure storage, vacuum storage, sub-atmospheric storage, etc. Reduction in pressure automatically reduces the partial pressures and availability of all components of air, notably the oxygen content. Hence the low pressure or vacuum is used to reduce the oxygen concentration in the atmosphere. For example, lowering the pressure to 1/10th of the atmospheric pressure automatically reduces the oxygen concentration to 2.1% from the 21% at atmospheric condition. It will also simultaneously reduce the N_2 concentration to 7.9% and humidity to below 10%. Since the air will be very dry, it will need to be humidified after the pressure reduction, which will bring back the moisture to the saturated level causing both O_2 and N_2 levels to go down a little further.

So hypobaric CA is a form of CA storage in which the produce is stored at sub-atmospheric pressures under the condition of a continuous flow of saturated air. Although vacuum cooling also holds the produce under sub-atmospheric conditions, the hypobaric set-up is different from the one normally used for vacuum cooling in many ways:

1. It provides a continuous flow condition (no flow in vacuum cooling).
2. It provides saturated air (highly unsaturated in vacuum cooling to promote transpiration).
3. There is no cooling in the hypobaric system unless intentionally induced at the beginning because of the saturated water vapor conditions.
4. It requires an indirect contact refrigeration system while vacuum cooling does not require any.

The hypobaric system consists of:

1. a gas/air tight chamber,
2. a powerful vacuum pump,
3. a pressure regulator,
4. a flow regulator,
5. a humidifier and humidity controller and
6. an indirect (jacketed) refrigeration system and temperature control.

Characteristics of Hypobaric System (Low pressure storage, LPS)

1. Air Composition

The ideal gas law states that:

$$P_{atm} V_{atm} = P_{lps} V_{lps}$$

if

$$P_{lps} = (1/10) P_{atm}$$

then,

$$V_{lps} = 10 V_{atm}$$

That means at lower pressures the gas occupies a larger volume (lower density); the other way is true as well, i.e., at higher pressure the gas will be compressed and will occupy a lower volume (higher density).

The volume of a pressure chamber or hypobaric storage is fixed. Hence to create a lower pressure or vacuum, some of the existing gas must be removed and the rest allowed to expand to the full volume. Thus to reduce the pressure in the chamber to (1/2) atmosphere (50% vacuum), then 50% of the gas must be removed and the remaining 50% should be allowed to expand to occupy the whole volume. This will reduce the available concentration of all active components in the system in proportion with the reduction of pressure.

For example, dry air at atmospheric pressure has approximately 21% oxygen and 79% nitrogen excluding other minor components like carbon dioxide, ethylene, etc. Hence when the pressure is reduced to 1/2 atm., available oxygen concentration will be 10.50% and nitrogen 39.5%. However, when moisture is present as an additional component, these will slightly change. But it should be remembered that if the original relative humidity is 100% (saturated air), it will decline to 50%, unless the air is humidified after it is subjected to vacuum!

Example calculation:

Let us say we have air at 10°C and 100% RH to start with and we apply vacuum to reduce the pressure to 1/10 atm (10 kPa);

Atmospheric pressure of air is 1 atm or 101.3 kPa

Partial pressure of water vapor in air when saturated at 10°C is 1.2 kPa (steam tables). Assume this is part of the total pressure of air (with water pressure added): Pressure of air is then 100 kPa.

Outside (101 kPa)	In the LPS (10 kPa)
P_{H2O} is 1.2 kPa	P_{H2O} is 0.12 kPa (10%)
P_{O_2} is ~21% of 100 or 21 kPa	P_{O_2} is ~2.1 kPa (2.1%)
P_{N_2} is ~79% of 100 or 79 kPa	P_{N_2} is ~7.9 kPa (7.9%)

Assume that the air inside the chamber is now saturated with water vapor bringing the partial pressure of water vapor to its vapor pressure (1.2 kPa at 10°C). What happens to the composition?

Outside (101 kPa)	In the LPS (10 kPa)
P_{H2O} is 1.2 kPa	P_{H2O} is 1.2 kPa (10%)
P_{O_2} is 21 kPa	P_{O_2} is 21% (10–1.2) =1.8 kPa (1.8%)
P_{N_2} is 79 kPa	P_{N_2} is 79% (10–1.2) = 7 kPa (7%)

This means that the available oxygen concentration is reduced further from about 2.1% to 1.8% because it was partly replaced by water vapor.

The oxygen concentration in the atmosphere therefore depends on: chamber pressure, temperature, and relative humidity of the air inside the chamber. Such a reduction in oxygen concentration is also true in conventionally heated chambers. For example, in a steam sauna where the room can be pretty much saturated with water vapor, the available oxygen will be much lower than outside.

Example: What will be the oxygen concentration in a steam sauna at 70°C?

Water vapor pressure at 70°C is 31.2 kPa

Air pressure will be 101.3–31.2 or 70 kPa which is composed of oxygen and nitrogen in the 21:79% proportion. So the partial pressures are 31 kPa for water vapor, 14 kPa oxygen and 56 kPa for nitrogen. The relative percentages are then approximately 31% for water vapor, 14% for oxygen and 56% for nitrogen. At higher temperatures, this value will decrease further, reaching zero at 100°C.

2. Gas Diffusivity

The diffusivity of gases varies with pressure and temperature as given by the following model:

$$D = D_o \left[\frac{T}{T_o}\right]^n \left(\frac{P_o}{P}\right)$$

Where

D_o = Diffusivity at T_o (273K)
P_o = Normal pressure (101.3 kPa)
D = Diffusivity at pressure P and temperature T
n = a factor 1.75–2.0

At a low pressure system at (1/10) atm or 10 kPa and 10°C as in the earlier example (assuming $n = 2$):

$$D = D_o \left[\frac{283}{273}\right]^2 \left(\frac{1}{0.1}\right) \approx 11 D_o$$

Thus the diffusivity in the hypobaric system operating at 10 kPa will be over 10 times the diffusivity at regular pressure, as compared to a CA room. This will have a tremendous implication on the extension point of oxygen in the storage system.

3. Concentration Gradients

Fick's law states that

$$J = AD\left(\frac{\Delta C}{\Delta x}\right)$$

Therefore, J increases as A, D, ΔC increase and Δx decrease. For a given J and across a membrane of a given area (A) and thickness (x), D and ΔC are inversely related since their product must be maintained at the same level to give the specified J. Thus, D will become higher when ΔC is lower. Likewise, a higher D will result in lower ΔC.

Since $D_{lps} \sim 10\, D_{atm}$ the associated ΔC will reduce to 1/10 of the value at atmospheric pressure.

Normally, in conventional CA rooms, the oxygen concentration is maintained at about 2% to give the necessary flux for oxygen to reach the tissue to maintain aerobic respiratory activity. In hypobaric systems, oxygen can be reduced to almost 0.2% without the danger of imposing anaerobic respiration. This will obviously be desirable since it will reduce other oxygen-related changes in the system.

4. Oxygen Level Control

Oxygen control in CA systems is quite critical and elaborate. In a hypobaric system, oxygen can be directly controlled by controlling the system pressure (vacuum). Since pressure can be controlled easily at a ±1 kPa level, it means the oxygen concentration can also be easily controlled at the ±0.2% level.

5. Outward Diffusivity

Several gases have an outward diffusivity since they are produced in the product and are driven by the concentration gradient. This is true with respect to carbon dioxide, ethylene and other toxic volatiles. If these are not efficiently driven out, they will affect the quality and shelf-life of the product. Since the diffusivity of gases is enhanced enormously in the LPS system, the tissue will be almost devoid of any toxic gases. This will provide a very low ΔC for these gases. As soon as gas is formed it is pushed out of the produce and then driven out of the system. Thus, harmful gases including CO_2 and ethylene, are effectively flushed out of the system.

6. Relative Humidity

Hypobaric systems provide a very high relative humidity since air in them is brought to a saturation point after the pressure is lowered.

7. Good Temperature Control

The hypobaric system works on jacketed refrigeration for indirect cooling of room air. Since there is no direct contact with the refrigeration system, the system humidity is better maintained, which results in low respiration rate and low heat of respiration.

Heat of respiration (BTU/ton.day)

Lettuce	0°C	4320 (in air)	550 (10 mm)
Apple	0°C	1200 (in air)	250 (50 mm)
Asparagus	0°C	6720 (in air)	22200 (10 mm)

8. No Need for Adding CO_2

In conventional CA, CO_2 is added for two reasons: (1) to suppress the respiration rate and (2) to suppress the formation and activity of ethylene. Carbon dioxide is not needed in the LPS system because of the outward diffusivity of gases and the use of super low levels of oxygen.

9. Lasting Effects

Beneficial effects last long after the removal of produce from storage, a result of the "pure" state of the plant tissue devoid of toxic gases.

10. System Permits Good Air Circulation: No Fans Required

Typical Operating Conditions for Hypobaric Storage

Pressure:	1–50 kPa
Temperature:	–2 to 15°C
RH:	80–100%
Air speed:	3–4 air changes/day

Disadvantages of Hypobaric System

1. Expensive.
2. Special structures needed. Not easy to convert existing structures.
3. Some stored products may lack flavor.
4. Maintaining humidity can be a problem when leaks occur.
5. Potential for inward implosion. Insurance implications.

CHAPTER 11

Packaging of Fruits and Vegetables

INTRODUCTION

PACKAGING of fruits and vegetables provides one of the most important functions—protecting the contents during storage, transportation and distribution against deterioration, which may be due to physical, chemical or biological causes. Packaging can be applied in the field, during production, or via subsequent handling in packing houses, processing plants or distribution centers. Packaging forms the last link between the produce and the consumer and plays an important role in the safe delivery of the produce. Robertson (1992) defined packaging as "the enclosure of products, items or packages in a wrapped pouch, bag, box, cup, tray, can, tube, bottle or other container to perform the following functions: containment; protection; and/or preservation; communication; and utility or performance." There has been a tremendous growth in development and application of packaging technologies for foods processed by both traditional as well as new food processing technologies. This has been driven by evolution of new packaging materials, package designs, package functions, sophistication in distribution and marketing techniques, consumer demand for convenience and unitized products and increasing production, processing, handling and energy costs.

For fresh fruits and vegetables, packaging is one of the most important steps in the handling and marketing of the produce from the farm gate to the consumer. There are different kinds of packages—bags, crates, baskets, cartons, bins of all sizes and shapes. These also are made from varied materials like wood, paperboard, plastic, metal, etc. The diversity of the packages and packaging materials for produce alone includes more than 1000 different types of packages with scores of different materials in North America alone. The numbers appear to increase as the industry introduces new packaging materials and new concepts, while many serious and fruitful discussions have been held about standardizing containers, container sizes and pallets. Although the packaging industry generally agrees container standardization is the right way to reduce package and transportation costs and improve handling efficiency, the trend over the years indicates that new package type and sizes are continually being introduced to accommodate the diverse needs of wholesalers, consumers and processing operations.

According to Wills *et al.* (1989), modern packaging must comply with the following requirements:

1. The package must have sufficient mechanical strength to protect the contents during handling, transport and stacking.
2. The packaging material must be free of chemical substances that could transfer to the produce and become toxic to humans.
3. The package must meet handling and marketing requirements in terms of weight, size and shape.
4. The package should allow rapid cooling of the contents. Furthermore, the permeability of plastic films to respiratory gases also could be important.
5. Mechanical strength of the package should be largely unaffected by moisture content (when wet) or high humidity conditions.
6. The security of the package or ease of opening and closing might be important in some marketing situations.
7. The package must either exclude light or be transparent.
8. The package should be appropriate for retail presentations.
9. The package should be designed for ease of disposal, re-use or recycling.
10. Cost of the package in relation to value and the extent of contents protection required should be as low as possible.

PACKAGE AS A HANDLING UNIT

The first function of the package is to contain the fruits and vegetables after the harvest and then to act as a handling unit. The many different types of harvest containers range from small bags or baskets to larger wooden or plastic containers, depending on hand or mechanized harvesting operation. Additionally, the containers will be used to handle the product. The container must enclose the produce in convenient units for handling and distribution. As a handling unit, it serves to:

- Carry the commodity from the field to the packing house (bulk containers).
- Move the product within the storage facility (bulk containers).
- Move the product through the distribution and marketing chain (shipping and retail containers).

For transport after purchase the package can segregate the products. Multi-wall bags like the ones used for onions and potatoes also can serve as a self-containing home storage unit until a product is consumed. These can serve some additional functions with respect to minimizing transpiration loss, protecting from exposure to light as well as creating modified atmospheres within the package. Figure 11.1 shows examples of commercial packages/containers used as handling units.

Further, a package helps to unitize the product and facilitates unitized

FIGURE 11.1. Some commercially diverse container types as handling units.

handling. Container unitization and standardization are important aspects of storage and transportation. Individual containers may be consumer units which are small retail containers varying from a few hundred grams to a few kilograms. Shipping containers generally accommodate more product; for example, up to 10 to 20 kg for manual handling and up to 250 kg for bulk handling by using forklifts. There are more than one thousand types of containers with hundreds of different dimensions, used for more than 50 common commodities.

In the international trade, container standardization has been recognized as a necessity. Container and pallet size standardization are being considered seriously to help in international trade. They reduce cost of handling and improve the efficiency of loading. Most pallet racks and automated pallet handling equipment are designed for standard-size pallets. They make efficient use of truck and van space and can accommodate heavier loads and more stress. The adoption of a pallet standard throughout the produce industry would also aid efforts toward standardization of produce containers.

A standard pallet is generally 100 × 120 cm. Shipping containers are designed to maximize the pallet space. Four common sizes suggested by OECD (Organization of Economic Cooperation & Development) are: 40 cm × 30 cm; 50 cm × 30 cm; 50 cm × 40 cm and 60 cm × 40 cm. Note that all these fit the standard pallet, but space efficiency is maximum with the middle two, unless different arrangements are made by mixing two or three different sizes. Figure 11.2 illustrates the different combinations that have 100% space efficiency.

PROTECTION FROM PHYSICAL AND MECHANICAL INJURIES

Protection is the second major function of a package. The package has to offer the produce protection from various physical & mechanical injuries, such as cuts, compression, impact and vibration.

Cuts and punctures: These injury result from sharp edges on handling equipment, presence of nails and rough edges on packages resulting in rupturing of produce skin. This is predominant during harvesting and field transportation.

Compression injury: This results from the compression of produce (squeezing) due to produce weight within the package (bulk containers) and packages stacked one above the other causing over-stacking, which may be compounded by overfilling. Underfilling in such over-stacked

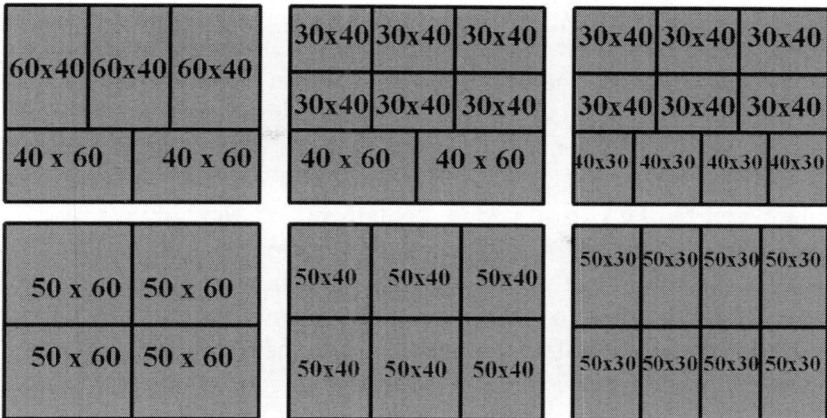

FIGURE 11.2. Full space efficiency of different containers on a standard pallet with different arrangements.

situations can cause package collapse. The main reasons for compression damage are therefore inadequate package or improper packaging.

Impact injury: This results from the sudden application of force on produce. It can happen because of dropping of the produce from excessive heights during harvesting and handling, dropping of filled containers, dropping of heavy objects on produce, jerky movements during transportation and impact shocks (sudden breaks, bumpy roads).

Vibration or abrasion bruise: This is a common injury caused to fruits and vegetables during transportation. It is caused essentially by product movement inside the container resulting in abrasion damage from rubbing of the produce against each other or against the container surface.

Produce types differ with respect to their sensitivity to the different kinds of mechanical damages. For example, apple, ripe banana, cantaloupe, peach, strawberry and tomato are highly susceptible to compression injury; apple, banana, peach and tomato also are susceptible to impact injury, while apricot, banana, grape, nectarine, peach, plum and squash are very susceptible to vibration injury. Grape, pear and plum are quite resistant to compression injury. Figure 11.3 illustrates the different mechanical injuries caused in the postharvest chain.

General Symptoms of Mechanical Injuries

The following are the general symptoms of the different mechanical injuries: surface and internal discoloration, loss of appearance and decreased market value. All form avenues for the spread of infections, increased respiration and chemical or enzymatic activity, eventually resulting in accelerated spoilage and diminished market value.

Good handling practices are the best way of minimizing the impact of the different mechanical injuries. Cuts and similar bruises can be minimized by the use of smooth containers and handling equipment. Compression damage can be controlled by using correct filling weights, using containers of adequate strength for containing the produce and by isolating the pressure of stacked containers onto the produce. Impact damage can be minimized by reducing drop height during harvesting and handling of the produce; cushioning of all contact points; avoiding falling of the produce and jerky movements of transport vehicles; and by wrapping the produce with cushioning materials. Vibration damage can be largely eliminated by restricting and preventing produce movement within the container during transportation and handling. This is commonly done using restrainers, individual wrapping and cushioning. Figure 11.4 shows some well-designed plastic containers that can be used for harvesting instead of wooden containers, as well as several mechanisms used for preventing damage to packaged fruits and vegetables.

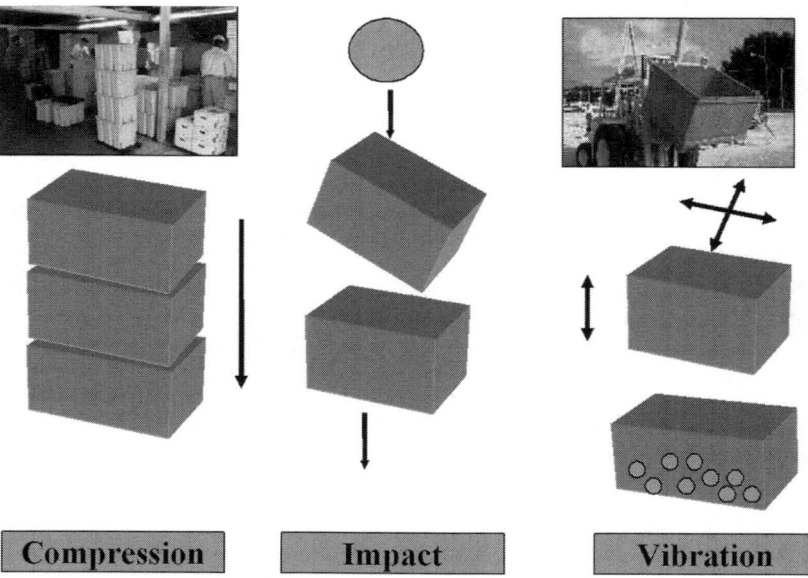

FIGURE 11.3. Illustration of different types of mechanical injuries.

FIGURE 11.4. Well designed harvest containers and packages and cushioning materials used to prevent mechanical damage in shipping containers.

PROTECTION FROM MOISTURE LOSS

Packaging helps to reduce moisture and weight loss by providing a gas and moisture barrier. While a moisture barrier is helpful, the gas barrier property needs to be appropriately controlled as in the case of modified atmosphere storage, otherwise it might create adverse conditions. However, if used appropriately, it can reduce transpiration losses and at the same time reduce the rate of respiration thereby helping to extend shelflife. When choosing the appropriate packaging material, the following factors should be considered: gas barrier properties, moisture barrier properties, anti-fog properties, machinability, mechanical strength, sealability and performance vs cost. One of the most important characteristics is their barrier properties to oxygen, carbon dioxide and water vapor, which varies greatly from material to material.

The barrier properties to oxygen, carbon dioxide and moisture also may be different for each material and also vary as a function of the relative humidity and temperature of the storage conditions. When gas barrier properties cannot be controlled, the package should be provided with appropriate vents to permit gas exchange. It also conserves the produce moisture within the package and hence minimizes transpiration. Such protection may be provided by wax coating of containers or boxes, protective wrappings within containers, plastic bags with holes for gas exchange, box liners, etc. In the case of bulk containers or crates in the open format, placing them under shade, cool places and occasional sprinkling with water may be helpful like, as is done in supermarket display units. Examples of packages that help reduce moisture loss are shown in Figure 11.5.

PROVIDING A SANITARY ENVIRONMENT

A package provides a sanitary environment and protects produce from extraneous contamination. Without the package, the produce may be exposed to an environment damaging to the produce, for example, plant pathogens, spoilage bacteria, chemicals and conditions that may accelerate deterioration, such as light, high or low humidity, dust, etc. Many of these may seem unimportant at the outset, but can certainly form avenues of deterioration.

FIGURE 11.5. Examples of packaging that give protection against moisture loss.

FIGURE 11.6. Example of a package with vent holes that facilitates ethylene treatment.

FACILITATING IMPORTANT TREATMENTS

Packaging influences the product temperature during cooling and storage. It can adversely affect the cooling process by causing significant delays since it can limit the contact of the cooling medium with the produce. Proper design of packaging is essential in order to accommodate cooling's effects. Since packages also can limit access to active components like ethylene during ripening treatments, they must be properly designed with appropriate vent holes for the penetration of ethylene into the package. Similar precaution must be taken during fumigation treatments for insect control. In general, large vents are better for all these applications; however, vents decrease package strength. Hence some compromise is essential between these two functions—protection vs treatment efficiency. Containers can provide advantage in focused application of certain treatments like cooling in forced air, hydro and vacuum and in package icing operations, or to facilitate ethylene ripening prior to retailing (Figure 11.6). It will certainly protect the product from direct exposure to air in cold storage and prevent/reduce transpiration.

PREVENTION OF PILFERAGE

The package will contain the product in an enclosure thereby preventing direct access of the product to consumers who may be tempted to test the quality of the product by squeezing, tasting, etc. It also will help to prevent product from spilling during handling.

IMPROVEMENT OF SALES PROMOTION

The consumer sees the package before seeing the product. So it can enhance sales appeal. An attractive package is the first step in attracting the consumer. Sales-related information can be included on the package to promote it. Recipes, nutritional labels and enticing pictures of the product are printed on packages to increase sales.

COMMUNICATION WITH CONSUMERS

Communication is important in today's marketing. The package provides an avenue to communicate to consumers information they like to see. That information can vary from product to product ranging from simple identification of the product to more complex data and advertisements (Figure 11.7). The package label generally is used to identify and provide useful information about the produce. It can provide information, such as the produce name, brand, size, grade, variety, net weight, count, grower, shipper and country of origin, etc. Today, most packages contain information on nutrients (mandatory for processed products), and in some cases useful recipes and other information specifically directed at the consumer. In consumer marketing, package appearance has become an important part of point of sale displays.

PROMOTION OF PRODUCT IDENTIFICATION AND COMPANY RECOGNITION

The packaged product is the entry of the company to the consumer market. Packages can include identification trademarks of the com-

FIGURE 11.7. Communication examples for selected fruits and vegetables.

FIGURE 11.8. Example of bar codes of some companies.

pany, of the product and of any other idea the manufacturer wants to communicate. Today, many top manufacturing companies have monopolized this issue. Many brands of products have been so closely associated with some companies that they are almost considered synonymous.

FACILITATION OF SALES

Imagine in today's market if the sales clerk had to look up the price information for the product and its weight to calculate the cost of each item. Such information is provided on the package using a machine readable language that can be combined with online weight or numbers to compute the cost. Universal Product Codes (UPC or bar codes) often are included as part of the labeling (Figure 11.8). The UPCs used in the food industry consist of a ten-digit machine readable code. The first five digits are a number assigned to the specific producer (packer or shipper) and the second five digits represent specific product information such as type of produce and size of package. These are integrated at the market level with the price information to facilitate easy checkout at sales counters.

GENERATION OF MODIFIED ATMOSPHERE WITHIN THE PACKAGE

Modified atmosphere is a very important function of today's custom designed packages and this aspect will be discussed in a later section. The permeability of the packaging material to oxygen and carbon dioxide will dictate its usability for produce packaging since the produce needs a continuous supply of oxygen for respiration and carbon dioxide, a product of respiration needs to be continuously removed from within in order to maintain the desirable aerobic respiration.

TYPES OF CONTAINERS

At and after harvest, fruits and vegetables are handled in different types of containers in the field and at various outlets. These are broadly classified into three groups: (1) Harvest and Field Containers, (2) Shipping and Storage Containers and (3) Retail and Consumer packs.

Harvesting and field containers: These are mostly used to collect and move the harvested produce to the packing house. There are many types depending on the crop, region and availability of materials. Mechanical harvesting systems often use bulk bins and pallet boxes, and the produce intended for processing are directly filled into open wagons and trailers. Containers come in different sizes and types and depend on the commodity type and handling practice:

- Baskets, buckets, pales of wood, metal and plastic.
- Boxes of wood, paperboard, plastic.
- Bins and bulk boxes of wood and plastic.
- Crates made of wirebound wood, plastic and wire.
- Bags of canvas and nylon mesh.
- Consumer packs of plastic and paper pulp.

Shipping containers: A shipping container is a handling unit used to facilitate moving of horticultural produce through transport and storage. Packaging for shipping and handling is engineered to be more functional in order to protect produce from bruising, vibration and the weight of other stacked containers. These containers are designed to be more sturdy to permit reasonable stacking without collapse or compression damage to the produce. Shipping should not detrimentally affect the respiratory exchange of O_2 and CO_2 into and out of the package and at the same time should allow the dissipation of the heat of respiration of fresh fruits and vegetables. The ideal pack consists of a compact-fill, with some cushioning material to protect and restrain the product, without a bulge on the lid and with the container having sufficient strength to withstand the stacking height. Common shipping containers include: nailed wooden boxes and crates, wirebound boxes and crates, plywood boxes and baskets, corrugated and multiwall paperboard containers. Fibreboard and paperboard (corrugated) cartons are becoming popular for

FIGURE 11.9. Diversity of containers used for produce packaging.

shipping both tropical and subtropical fruits. Their light weight and low cost are advantages.

Consumer Packages: Use of small consumer-sized packages for produce has grown with the increase in large self-service markets for retailing. It may consist of a paper or a plastic bag available for customers to select, package and weigh their purchases. A favorite type of consumer package for preparation at the wholesale level or in the retail store is the molded tray. Trays can be made of chipboard, molded foam plastic or clear plastic.

Consumer packages are of the following types: (1) bags made of paper, film or cotton or plastic mesh; (2) trays of molded pulp, paperboard, plastic or foamed plastic; (3) folding paperboard cartons, sometimes with a clear plastic window or with dividers for individual fruits; and (4) small rectangular or round baskets, coated or waxed paperboard, etc.

Figure 11.9 illustrates different types of packages used for produce at different stages—harvesting, shipping, handling, retailing, etc.

TYPES OF PACKAGING MATERIALS

A variety of packaging materials, with specific functional properties, are commercially available for packaging fresh fruits and vegetables. A useful website http://truckpallet.com/pallet-handling.php gives excellent details about the different types of packaging materials used for produce handling.

Wood

Wooden containers were traditionally used for the bulk transportation of fruits and vegetables to marketplace. Wooden containers generally are strong and offer good mechanical protection and stacking characteristics, but they are poor moisture and gas barriers. Wooden containers are gradually being replaced by polystyrene, polypropylene and polyethylene containers, which are lighter and have lower transportation costs.

Pallets and Pallet Bins

Pallets form the base on which most fresh produce is delivered to the consumer. Because many are of a non-standard size, pallets are traditionally wood based and are built as inexpensively as possible and discarded after a single use. The 100 cm × 120 cm (40" × 48") pallet has evolved over the years as the standard size for the pallet. Depending on the size of produce package, a single pallet may carry from 20 to more than 100 individual packages. Because these packages are often loosely stacked to allow for air circulation and often difficult to stack evenly, they must be secured (unitized) to prevent shifting during handling and transit. Plastic straps and tapes or plastic stretch film are widely used to secure produce packages. The plastic film helps to protect the packages from loss of moisture, makes the pallet more secure against pilferage, and can be applied using partial automation. The plastic film also can severely restrict proper ventilation. A common alternative to stretch film is plastic netting, which is much better for stabilizing some pallet loads, such as those that require forced-air cooling. Another low-cost alternative for pallet stabilization is the application of a small amount of special glue to the top of each package. As the packages are stacked, the glue secures all cartons together. Pallet bins are primarily made out of wood (milled lumber or plywood) and are used primarily to move produce from the field to the packing house. Capacities vary depending on the type of produce. Generally, these are boxes made with a pallet as the base and hence are designed around the standard pallet size of 100 cm × 120 cm, while generally the height varies depending on the product type. In some large operations double-width pallet bins are used, which have twice the width of the standard bin. Size standardization is essential for pallet bins since the filled containers may be subjected to forced-air cooling operations and also are used as bulk storage bins. In order to prevent bruising of produce, they need to be carefully manufactured to prevent protruding of nails, straps, etc. The life of pallet bins can vary between 5 and 10 years, depending on weather conditions.

Wire-bound and Wooden Crates, Baskets

Wooden wire-bound crates are used extensively for many vegetables that are commonly hydro-cooled. These are sturdy, rigid and have very high stacking strength and generally are water resistant. If necessary, they may be disassembled after use. Wooden crates were extensively used in the past, and they generally are heavy. They still are used for some commodities like apples and grapes. Wire-reinforced wood veneer baskets and hampers are durable and may be nested for efficient transport when empty. However, cost, disposal problems and difficulty in efficient palletization have severely limited their use to local grower

FIGURE 11.10. Pallets, pallet bins, wirebound boxes and crates (wood, paper and plastic).

markets, where they can be re-used many times. Examples of pallets, bins, crates, etc. are shown in Figure 11.10.

Paperboard

Corrugated fiberboard (or paperboard) is a relatively low cost, versatile and dominant container material used for transporation and handling fruits and vegetables. Double or triple corrugated fiberboard is the predominant form for produce containers. The multi-layer corrugated paper board is made by sandwiching layers of corrugated paperboards between plain paperboards. A three-layer board will have one corrugated board sandwitched between the inner and outer boards, while others will have additional corrugated board-plain board combinations. Since moisture resistance is important for produce handing, the inner and outer layers need to have special coating for making them moisture resistant, for example, a coating of wax. Multi-wall corrugations usually are made for heavy-duty shipping containers, especially bulk bins because they are required to maintain large loads and high stacking heights. In recent years, large double-walled or even triple-walled corrugated fiberboard containers have increasingly been used as pallet bins to ship bulk produce to processors and retailers. Cabbage, melons, potatoes, pumpkins and citrus produce have all been shipped successfully in these containers. Cold storage conditions and high humidities reduce the strength of fiberboard containers. Unless the container is specially treated, the absorbed moisture can significantly reduce the strength of the container. One of the major disadvantage of paper as a packaging material is its poor barrier properties against moisture, gases, grease and odors. Waxed fiberboard cartons are frequently used for produce that is either hydrocooled or package-iced.

Cold storage conditions (low temperature and high humidities) tend to reduce the strength of paperboard containers. Unless the container is specially treated, i.e., wax coated, moisture absorbed from the environment and the high moisture from fruits and vegetables can degrade the packaging material. Wax coating is the most commonly used technique to impart moisture resistance, but plastic coating also is employed. Unlike those used for durable goods, packages for fresh produce usually cannot carry much of the vertical load without damage to the produce. Therefore, one of the important characteristics of corrugated fiberboard containers is stacking strength. Creating vent holes to facilitate cooling and other treatment significantly reduces the strength of the package.

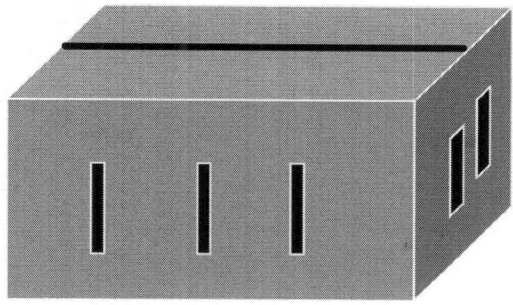

FIGURE 11.11. An illustrated package with designed side vent holes.

The majority of the stacking strength of corrugated containers is supported by the corners. Hence, the vent holes must be placed away from the corners and cover about 5% of the side surface area. The vertical cross section of the vent holes also should be minimized to limit package weakness. A horizontally placed long vent hole will result in poor package stacking strength. Vent holes on the top and bottom do not facilitate circulation because they are often blocked in stacked containers. A typical corrugated paper board container with vertical vent holes is illustrated in Figure 11.11.

Pulpboard Containers

Containers made from recycled paper pulp and a starch binder also are used for small fruits. These are available in different shapes and sizes and are relatively inexpensive. Pulp containers can absorb surface moisture from the product surface and prevent mold growth in moist produce. These containers also are biodegradable and recyclable.

Paper and Mesh Bags

Consumer packs of potatoes and onions are the main produce items now packed in paper bags. The more sturdy mesh bag has much wider use. In addition to potatoes and onions, cabbage, turnips, citrus and some specialty items are packed in mesh bags. In addition to its low cost, mesh has the advantage of uninhibited air flow. Good ventilation is particularly beneficial to onions. Use of bags as packages has several common disadvantages. Bags in general do not palletize well or efficiently fill the space inside shipping containers. They do not offer protection from rough handling. Mesh bags provide little protection from light or contaminants. In addition, produce packed in bags is perceived by the consumer to be less than the best grade.

Plastics

There has been a tremendous increase in the use of plastics replacing traditional packaging materials, such as glass, metal and paper. The raw materials for plastics are petroleum, natural gas and coal. They are formed by a polymerization method, which creates linkages between many small repeating chemical units (monomers) to form large molecules or polymers. Many plastics contain very small amounts of additives, such as plasticizers, antioxidant lubricants, antistatic agents, plus heat and U.V. stabilizers. These are added to facilitate processing of plastics or to impart desirable properties to the plastics. For example, plasticizers are added to soften plastics, thus making them more flexible and less brittle for use in cold climates or with frozen stored products. The advantages of plastics as packaging materials are: their cost is relatively low; have good barrier properties against moisture and gases; are heat sealable to prevent leakage of contents; are suitable for high-speed filling; have wet and dry strength; are suitable for printing; are easy to handle and convenient for the manufacturer, retailer and consumer; add little weight to the product; and fit closely to the shape of the food, thereby wasting little space during storage and distribution (Fellows, 1988).

Rigid Plastics

Referred to as clamshells, rigid plastics are gaining popularity because they are inexpensive, versatile, provide excellent protection to the produce and present a very pleasing consumer package. Clamshells are used extensively for several small size fresh fruits and vegetables and precut produce. Molded polystyrene and corrugated polystyrene containers have been test marketed as a substitute for waxed corrugated fiberboard. Heavy-molded polystyrene pallet bins have been adopted by number of growers as a substitute for wooden pallet bins. Although they are more expensive than wooden bins, they have a longer service life, are easier to clean, are recyclable, do not decay when wet, do not harbor disease, and may be nested and made collapsible. As environmental pressures continue to grow, the disposal and recyclability of packaging material of all kinds has become a very important issue. Common polyethylene may take several hundred years to break down in a landfill. The addition of 6 percent starch will reduce the time to 20 years or less making the use of these materials more practical.

Plastic Films

Plastics may be made as flexible films or as semi-rigid and rigid containers to meet the varying packaging and processing requirements of food. Plastic films are made with a wide range of mechanical, optical, heat sealable and barrier properties. Furthermore, they can be coated with another polymer or metallized to give a laminated structure with superior properties. Some common flexible films are: cellulose, polyethylene, polyester, polyamide, polypropylene, polystyrene, polyvinyl chloride, poly-vinylidene chloride, ethylene vinyl acetate, ethylene vinyl alcohol, ionomer, etc.

Plastic-film Bags

Numerous transparent and translucent plastic films of various compositions are available commercially, and some of them at lower cost than kraft paper, cotton cloth or burlap. Prefabricated bags are available from many manufacturers, and also may be fabricated at the user's premises by machines, that form and heat-seal bags from rolls of flat film. The advantages of plastic-film bags are good visibility, limited permeability to water vapour and good strength and tear resistance.

Mesh Bags

Mesh bags provide good ventilation to their contents. The netting with openings between strands allows free movement of air to and from the interior of the bag. Mesh bags can be fabricated from several materials like plastic strands, cotton thread as well as twisted strands of processed paper. The advantages of mesh bags are ventilation, prevention of humidity build up, good visibility of product and easy closure. Mesh bags are widely used for a variety of fruits and vegetables like apples, oranges, onions, corn, etc.

Shrink-film Wraps

Films of a number of types can be given heat-shrink characteristics

FIGURE 11.12. Shrink wrapping examples.

by stretching under controlled temperatures and tensions to form molecular orientation, after which the film is cooled in the stretched condition to maintain its form. Films such as polypropylene, polystyrene, polyethylene and rubber hydrochloride, can be converted to shrink-films by the molecular orientation method. After the shrink-film is applied to the filled trays in tubular or heat-sealed wrap form, the packages are passed through a heat tunnel to shrink the film cover. This immobilizes the fruits to reduce the possibility of physical damage during handling. Figure 11.12 shows some commercial products that are produced either by shrink wrapping the produce directly or on a tray along with a commercial shrinkwrap packaging machine.

PACKAGING REQUIREMENTS FOR FRUITS AND VEGETABLES

The shelf life of packaged fruits and vegetables is controlled by the perishability of the produce (susceptibility to enzymic or microbiological deterioration, mechanism of spoilage, and the requirement for or sensitivity to oxygen, light, carbon dioxide and moisture) and the properties of the package. Prevention of moisture loss is one of the most important factors that control the shelf life of fruits and vegetables. Fruits and vegetables are high-moisture products with moisture contents ranging from 75–95%. Transpiration moisture loss leads to product shrivilling, shrinkage, wilting, loss of quality and salability. The package must be a barrier to water vapour, in order prevent moisture loss from the product. However, they cannot be barriers to oxygen and carbon dioxide in order not to compromise produce respiration. The use of small perforations in some films to ensure a constant supply of oxygen generally has no appreciable effect on moisture loss since the package helps to create a high humidity environment within the package, which is adequate to suppress the respiration.

Fruits and vegetables are living tissues and continue to respire and transpire even after harvesting. The desired aerobic respiration requires oxygen, and as a result of respiration, carbon dioxide is produced. There must be adequate oxgen for aerobic respiration to continue and further, the resulting carbon dioxide must be removed from the package. Therefore, packaging materials for fruits and vegetables should be permeable to both oxygen and carbon dioxide. Holes are provided in many plastic bags for this purpose since most plastic bags are not adequately permeable to these gases.

The package should not adversely affect the various treatments, such as cooling, cold storage, ripening, fumigation, etc. The package should retain desirable flavor volatiles from the fresh produce and prevent the pickup of off-odors from plasticizers, printing inks, adhesives or solvents used in the manufacture of the packaging material.

Packages must protect the produce from mechanical damage caused by transportation or handling e.g., cuts, vibration, compression and impact damage. Proper precautions must be taken by cushioning, restraining and properly packing in containers of adequate stack strength.

MODIFIED ATMOSPHERE PACKAGING

Modified Atmosphere Packaging (MAP) can be defined as "the enclosure of food products in a barrier film in which the gaseous environment has been changed or modified to slow respiration rates, reduce microbiological growth and retard enzymatic spoilage with the intent of extending shelf life" (Young *et al.*, 1988). MAP is an increasingly popular method for shelf life extension of food products when an extended shelf life at refrigerated temperatures is required for products. It provides modified atmospheres inside the package, which can be beneficial to the produce for extending the shelf-life. The commercial success of controlled and modified atmosphere storage provides the basis for MAP's popularity. MAP represents a downsized version of a membrane system for CA storage. Like membrane storage stystems, MAP products are based on selective permeability of packaging materials for matching produce respiratory needs with the gases permeating across the packaging material. The atmospheric conditions in MA packages are more variable, less precise and relatively inefficiently controlled. Thererore, MAP should not be considered as a replacement for CA, but it can be a supplement. These packages also can provide a smooth transition for products removed from CA rooms to prevent their sudden exposure to atmospheric conditions. A schematic demonstrating the extension of shelf life as a result of MAP is illustrated in Figure 11.13. Advantages of MAP are similar to those discussed for CA/MA storage in the previous chapter.

Principles of MAP

The proper atmosphere inside the package is achieved and maintained by selective diffusion of oxygen (into the package) and carbon dioxide (out of package). It must be recognized that the produce's respiratory process requires oxygen and therefore the produce continu-

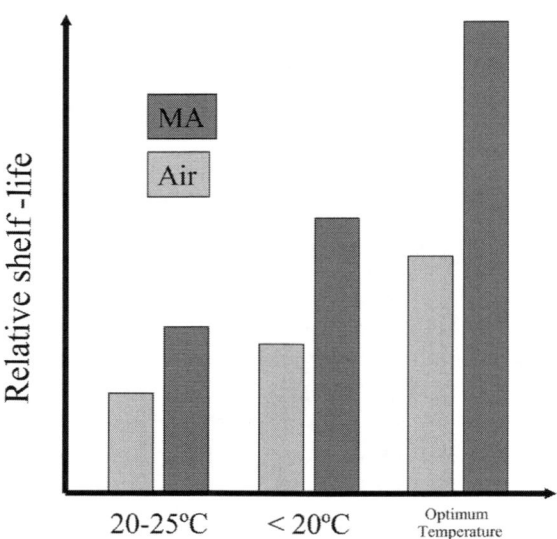

FIGURE 11.13. Relative shelf-life of produce under modified atmosphere storage.

ously consumes and depletes the oxygen present in the package. Thus, oxygen must be supplied into the package at the rate consumed by the produce respiration. Likewise, respiration results in the liberation of carbon dioxide which must be continuously removed from the package at the same rate at which it is produced as a result of respiratory activity. If these two requirements are not met, anaerobic respiration sets in relatively quickly as a result of oxygen depletion and carbon dioxide accumulation. Transfer of these gases into or out of the package depends on the permeability the packaging material to the respective gases and their concentration gradients. So creation of MAP is a very dynamic process, and its equilibrium represents a balance between produce respiration and package permeation. The respiratory demand depends on the nature, quantity and physiological state of the produce, temperature, atmospheric composition, presence of ethylene as well as mechanical and other stresses. The permeability of the package, on the other hand, depends on the nature of the packaging material, composition, density, physical structure and, of course, temperature and humidity conditions.

Just like CA/MA, MAP is a two-step process involving the generation of MA and then maintaining it. While the CA generation in commercial storage systems requires elaborate considerations, MA generation involves just two simple steps.

MA Generation

Two processes can be used to create the atmosphere within the packaged products. These are: (1) passive process and (2) active process.

Passive process (produce respiration). In the passive process, the atmosphere is generated as a result of the produce's respiration. The respiratory process results in O_2 consumption and liberation of CO_2. Hence, the oxygen concentration will slowly decrease to the desired level while the CO_2 will simultaneously increase. Since this process depends on produce respiration, it is a relatively slow process and therefore the benefits of MAP are not realized immediately.

Active process (gas flush). In the active process, the package headspace is flushed with a gas mixture of known and predetermined concentrations of O_2, CO_2 and N_2 for a quick control of respiration and of ethylene. Since the active process initially subjects the produce to vacuum followed by a gas flush, the vacuum also helps to remove toxic gases. A form-fill seal type of packaging system can be used for this

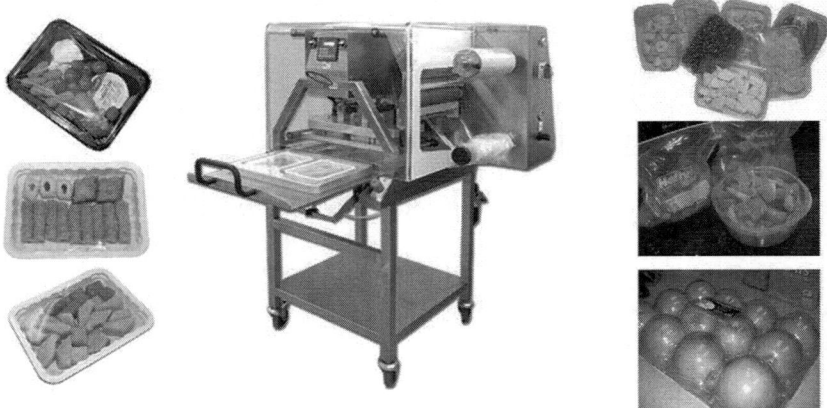

FIGURE 11.14. A commercial modified atmosphere (vacuum/gas flush) packaging machine and some examples of MAP products.

purpose or alternately the vacuum/gas flush packaging system used for semi-rigid can easily be adapted (Figure 11.14).

MA Maintenance

Maintaining the modified atmosphere (MA) is essentially done through the selective permeability of gases across the packaging material. In order to arrive at the optimal conditions inside the package, the respiratory requirements of product and permeability characteristics of the package need to be carefully coordinated.

There are many situations in postharvest handling of fruits and vegetables when modified atmospheres are created within the packages even when they are not intended. Examples are: use of air-tight containers or cold storage, packaging of produce in film wraps or bags, use of polyethylene liners in shipping containers, use of pallet shrouds, manipulation of shipping container vents, application of waxes or other surface coatings or use of plastic cover with diffusion windows. Such situations must be monitored carefully to make sure that damaging conditions do not result from a specific technique.

Influence of Film Permeability on MAP

Selection of a film with the correct permeability to oxygen and carbon dioxide is critical to the success of MAP for fruits and vegetables.

If the film is too permeable to oxygen, the product will respire as in outside air and there will be no advantage of MAP. On the other hand, if the permeability is too low, anaerobic conditions will soon be reached and the product will spoil and ferment.

Consider the following example as indicated by the initial and desired conditions of MAP with 5% O_2 and 10% CO_2 (Figure 11.15). In the passive system when the produce is packaged inside the container, the concentration of O_2 and CO_2 inside and outside will be same and hence there will be no concentration gradient for either of the two gases. However, in the active system, since the MAP conditions are instantaneously produced by gas flush and sealed, the 5% O_2 and 10% CO_2 is created inside while the outside air would have 21% O_2 and 0% CO_2 and hence the concentration gradients will be 16% for O_2 and 10% CO_2.

Now let us consider different scenarios like the film having no or equal permeability to either oxygen or carbon dioxide, as well as having a desired selective permeability to maintain the MAP conditions. Let us also assume the respiratory demands are met as far as possible and that we are considering a simple case of sugar for respiration (RQ = 1.0).

Let us discuss the different cases and try to assess the possible equilibrium conditions inside the package. In all cases, outside air is normal air with 21% O_2 and 0% CO_2.

An Example of Initial MAP Conditions

	Inside	Outside	ΔC
Passive	21% O_2	21% O_2	0%
	0% CO_2	0% CO_2	0%
Active	5% O_2	21% O_2	16%
	10% CO_2	0% CO_2	10%

Desired conditions: 5% O_2 10% CO_2

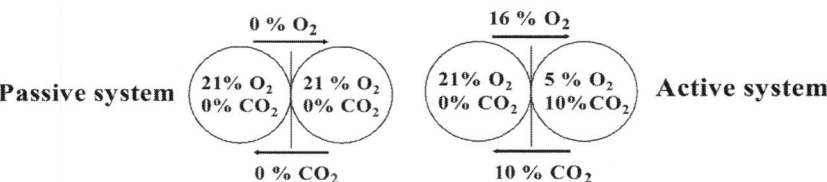

FIGURE 11.15. Example of intitial conditions for MAP (5% O_2 and 10% CO_2).

Case 1. Film with No Permeability

No-permeability film is a simple case with obviously damaging results. These packaging materials will be barriers to O_2 and CO_2. The oxygen present in the container, relatively at a higher level in the passive process, will continuously diminish with simultaneous accumulation of carbon dioxide reaching anaerobic conditions soon and the product will ferment and spoil. In the gas-flushed active process, the available oxygen is even lower and will reach the end stage even more rapidly. Hence, this is not a desirable situation.

Case 2. Film with Equal Permeability

The equal permeability situation produces two interesting possibilities as indicated in Figure 11.16. Passive and active processes of MAP creation yield different results.

Equal permeability implies presence of large pores, which will not discriminate between the two gases in question (O_2 and CO_2). This also can be achieved by creating carefully crafted perforations in barrier films to allow both oxygen and carbon dioxide to enter/exit.

In the passive system, there is no initial gradient for either O_2 and CO_2, since the respiratory process is under $RQ = 1$, which means O_2 consumed = CO_2 produced. Hence for every molecule of O_2 consumed, there is one molecule of CO_2 produced. If the required amount of O_2 comes through the package, then the same amout of CO_2 goes out. This means there will be no change in O_2 and CO_2 concentrations as a result of respiration. The final concentrations will be same as the initial concentrations.

However, in the active process, the scenario is different. Initially, there is a 16% concentration gradient for O_2 and 10% gradient for CO_2. So even though the permeability of the packaging material to these two gases is same, it will allow relatively more oxygen to come in and restrict the carbon dioxide going out. This means that with the passage of time, the oxygen concentration inside the package will tend to increase from the 5% level. Likewise, the carbon dioxide level also tends to increase beyond the 10% level due to the restriction imposed by the concentration gradient. However, this situation is highly transient. The increase in O_2 concentration will progressively reduce the ΔC for O_2, and the accumulation of CO_2 will increase the ΔC for CO_2. These two will equilibrate eventually at the middle level of concentration differ-

Equilibrium Conditions Inside the Package

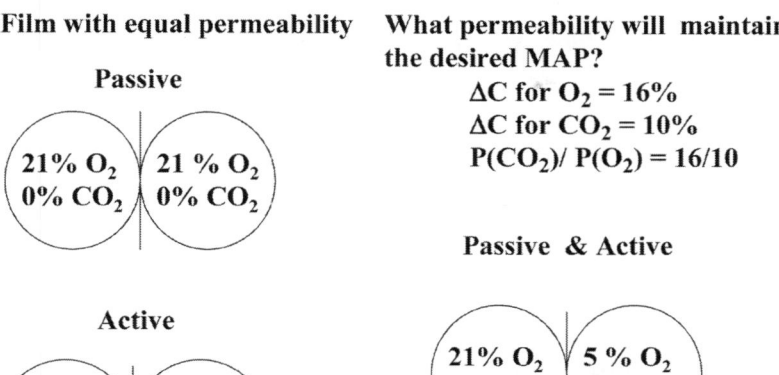

FIGURE 11.16. Equilibrium conditions achieved for MAP.

ence, which will be the average of the initial ΔCs for O_2 and CO_2. In this example, it will be 16 + 10 divided by 2 or 13%. Hence, the equilibrium condition will have 8% O_2 and 13% CO_2. This level of CO_2 may not suit all produce; the initial conditions can be chosen appropriately to give the desired CO_2 level. But a lower CO_2 concentration basically means higher O_2 since the sum of the two should be 21% (see above example, 8 + 13 = 21), which may reduce the effectiveness of the MAP.

Case 3. Selective Permeability for O_2 and CO_2

The discussion of Cases 1 and 2 clearly indicates that (1) films with no permeability don't find any use as packaging material for MAP of produce and (2) film with equal permeability cannot be used to create specific conditions since the concentrations of O_2 and CO_2 will have to be complementary, totaling 21% together. The only way to have a desired combination of O_2 and CO_2 inside the package is by having selective permeability to O_2 and CO_2. What should these be? Only two influencing factors are being discussed: concentration gradient (ΔC) and permeability (P). Higher concentration gradients and higher permeability will both allow ingress and egress of larger amounts of the respective gases. The gas flux (J) depends on the product of ΔC and P. In this dynamic scenario, they are going in the opposite directions. Hence to have the same J for both O_2 and CO_2 their $\Delta C \times P$ values need to be the same. In other words:

$$J = \Delta C(O_2) \times P(O_2) = \Delta C(CO_2) \times P(CO_2)$$

or

$$\Delta C(O_2)/\Delta C(CO_2) = P(CO_2)/P(O_2)$$

Hence in the above case, since $\Delta C(O_2)/\Delta C(CO_2)$ is 16/10, the permeability ratio $P(CO_2)/P(O_2)$ should be kept the same 16/10. In other words, the permeability to CO_2 should be 1.6 times the permeability for O_2 in order to offset the 1.6 times higher concentration gradient present for O_2 over CO_2. When such a package is designed on this concept, the resulting MAP conditions will be maintained at the desired level. This will be true for both active and passive cases although in the latter case, gas levels will go through a gradual change as in Case 2.

The MAP requirements for different fruits and vegetables vary. In general, the oxygen level for most temperate fruits like apples, pear, peach, apricot, etc., is reduced to about 2–3% and carbon dioxide levels are elevated to about 5% with a temperature set between 0 and 5°C. But some fruits like strawberry and cantaloupe can take 10–15% carbon dioxide while common tropical fruits like mango, papaya, pineapple and grapefruit tolerate up to 10% with CO_2 (optimum storage temperature 10–15°C). On the other hand, for some fruits like honeydew melon and tomato, carbon dioxide must be eliminated from the container since they can be damaged by its presence. Among the vegetables, most do well at a reduced oxygen level of 2–3%, while some, like asparagus, mushroom and spinach require storage at high oxygen (air). Generally, vegetables do well with 3–7% carbon dioxide, except that cucumber, lettuce and bell peppers do not tolerate carbon dioxide. Corn, beans and asparagus, avocado, broccoli, corn, mushrooms and spinach can tolerate from 10–20% CO_2 (Kader, 1986).

Assuming a general range of 2–3% O_2 and about 5% CO_2, the $\Delta C(O_2)/\Delta C(CO_2)$ ratio will be around 3–4, and hence the $P(CO_2)/P(O_2)$ ratio needs to be in the same range in order to provide the desired MAP. There are several films like special grades of polyethylene (especially low density PE), polyvinyl chloride, polyester (PET), polystyrene (PS) and polypropylene (PP) that have the ratio in the desired range. Among these, LDPE, PP and PS have relatively high permeability to both oxygen and carbon dioxide to be able to be successfully used for MAP. PET has a fairly low permeability to both oxygen and carbon dioxide.

Edible Films

Coating of fruits and vegetables with edible materials to preserve their quality and extend their shelf life has been in practice for centuries. The most common form of coatings of fruits and vegetables is wax to retard respiration, dehydration and senescence. Hot-melt waxes and carnauba oil-in-water emulsions have been used effectively for citrus fruits, apples, tomatoes and eggplants (Kester and Fennema, 1986).

Interest in edible films has intensified over the past few years due to the increased consumer demand for fresh, frozen and fabricated foods and consumer concerns about the environment. The most important characteristics of edible films are: (1) they are good gas and moisture barriers and (2) they may be suitable for coating with additives such as antimicrobials, antioxidants, nutrients and coloring agents.

The edible films most commonly used are derived from polysaccharides, proteins and lipids. Polysaccharide films can be made from starch, dextrins and cellulose derivatives. Protein films are made from collagen, gelatin, wheat and corn gluten and zein. Lipid films may be made from natural waxes and surfactants (Kester and Fennema, 1986). Lipid (hydrophobic) films are good moisture barriers. Fungicides can be applied to these films to retard yeast and mold spoilage. However, the coating must not be too impermeable a gas barrier so as to reduce anaerobic respiration causing physiological disorders (Kester and Fennema, 1986). Other materials that have been used as edible film coatings include: alginate, gelatin, pectin, acetylated monoglyceride, chitin and chitosan.

Edible coatings serve as carriers of food additives such as anti-browning agents, anti-microbials, colorants, flavors, nutrients, and spices improving shelf life and food quality, as they are selective barriers to moisture transfer, oxygen uptake, lipid oxidation and the loss of volatile aromas and flavors. The potential uses of edible coatings are to produce a modified atmosphere in the fruit thereby reducing decay, water loss, aroma loss, and exchange of humidity between fruits. Coatings delay ripening, color changes and improve appearance. The edible coatings should be stable under high relative humidity and are generally recognized as safe (GRAS). They should be colorless, tasteless and should have good water barrier, gas barrier and mechanical properties (Undurraga et al., 1995). Freshly harvested fruits and vegetables like mango, strawberries, peaches, tomatoes and papaya, which are subjected to faster deterioration and foods susceptible to the oxidation are good candidates for edible coatings.

Edible film components are divided into several categories, hydrocolloids, lipids, composites and plasticizers. Hydrocolloids include proteins, cellulose and its derivatives like alginates, pectins, starches and other polysaccharides. These are used where control of water vapour migration is not the prime objective but a barrier to oxygen and carbon dioxide is. These films are used mainly for improving the structural integrity of the packaging material of foods because of their good mechanical properties. Hydrocolloid films under the action of heat will dissolve, thereby not altering the sensory properties of the food. They are classified into film-forming carbohydrates and film-forming proteins. Film-forming carbohydrates include starches, plant gums, chemically modified starches, while film-forming proteins include casein, soy protein, whey protein, wheat gluten and zein.

Lipids include waxes, acyl glycerols and fatty acids. The lipid films are mainly used for making foods resistant to water vapour. Their other applications include adding gloss to the confectionery products. Their application as pure films is advised due to the lack of structural integrity and durability. Lipid films generally are used to retard respiration and reduce the loss of moisture.

Composite films constitute both lipid and hydrocolloid films. They are combinations of both hydrocolloid films and lipid films and have the advantages of both films, i.e., mechanical properties as well as barrier properties to water vapour and other gases. Addition of plasticizers in minor amounts to the edible film will alter the film properties, e.g., functional, organoleptic, nutritional, and mechanical properties. Addition of plasticizers, such as glycerol, acetylated monoglyceride, polyethylene glycol and sucrose, will improve the mechanical properties of the film. Other categories of plasticizers include antimicrobial agents, vitamins, antioxidants, flavors and pigments.

CHAPTER 12

Irradiation of Fruits and Vegetables

INTRODUCTION

FOOD irradiation is the process of exposing food to a controlled amount of energy called "ionizing radiation." We are constantly exposed to solar radiation and food is commonly irradiated when cooked in a conventional oven or a microwave. Infrared radiation is routinely used in food drying applications. Radiations that have enough energy to move atoms in a molecule or cause them to vibrate, but not enough to remove electrons, are referred to as "non-ionizing radiations." Ionizing radiation does have enough energy to remove tightly bound electrons from atoms, thus creating ions. The term "food irradiation" is only used to describe a process where a food is exposed to very high-intensity ionizing radiation utilizing gamma rays emitted by Cobalt-60 or Cesium-137 radioisotopes, machine-generated X-rays up to 5 MeV, or electron beams (EB) accelerated to acquire an energy level up to 10 MeV. The electromagnetic radiations of Cobalt-60 and Cesium-137 sources and the X-rays have good penetrability, while the accelerated electron beam has a relatively low penetrability. The advantages of X-ray and EB are that they can be shut down when not needed, while radioactive sources have to burn themselves out to ground state, which can take a long time. At the levels employed, none of these energy sources induce radioactivity in the food or its packaging, and the treatment has many technologically and technically feasible applications including significantly improving microbiological safety and/or storage stability of foods.

Food irradiation has decades of history and was developed as a scientifically established technology and safe food process during the second half of the 20th century. Irradiation technology has, however, never realized the technological breakthrough and commercialization of other common food processes like canning, freezing, dehydration, etc. Much of this is due to controversies associated with installation and process safety issues as well as alleged effects of irradiation on food components. Consumer and environmental lobby groups have been very vocal against the use of this technology; hence, in spite of its many scientific merits, irradiation has always looked like a "rebel" technology forced on consumers by opportunistic processors, and the technology has always been at the mercy of political policy.

Food irradiation at low doses delays ripening, inhibits sprouting and extends shelf-life by reducing spoilage organisms in fruits and vegetables, helping to meet quarantine standards for export to foreign markets. Slightly higher doses are effective for the disinfestation of insects in food—eliminating the risk of introducing foreign insects to a country. Higher doses significantly reduce or kill pathogens such as E.coli, listeria and salmonella in seafood, meats, and poultry, substantially improving the safety of the food. While food irradiation is not yet widely adopted, it has been declared "safe and wholesome" by the World Health Organization (WHO) and the U.S. Food and Drug Administration (FDA). Some 40 countries around the world also have approved the use of irradiation for more than 100 food types. More than 175 million pounds of spices and more than 15 million pounds of ground beef and poultry are irradiated each year in the U.S. And with recent legislation in the U.S. approving food irradiation as a quarantine treatment for certain fruits being imported into the U.S., food irradiation may be poised for substantial growth. The USA, South Africa, Netherlands, Thailand and France are among the leaders in adopting this technology. Currently regulations on food irradiation in the European Union are not fully harmonised. Some countries, such as Belgium, France, Netherlands and UK allow herbs and other foods to be irradiated, whereas other countries, such as Denmark, Germany and Luxembourg remain opposed. Regulations across the world make provision for labeling to ensure that consumers are fully informed whether foods or ingredients have been irradiated. Food irradiation is slowly gaining consumer acceptance in the U.S. and several other countries, but it is slow to gain support within many parts of Europe, including the UK, where the Food Standards Agency (FSA) recommends no extension of application. Many consumers are initially hostile to irradiation, but when the process is explained to them they become generally more in favor. There is a role for respected professional bodies to inform consumers of the advantages and limitations of the technology so that they can make informed decisions on buying and eating irradiated foods (IFST 2006).

HISTORICAL PERSPECTIVES

Food irradiation is not something new. However, it is being used more often and as a result is being more closely examined as a public health issue. Research on food irradiation began as early as 1905. Table 12.1 presents a chronological summary of major events in the progress and use of irradiation as reported by USEPA (US Environmental Protection Agency, Table 12.1). Some additional milestones are given in Table 12.2.

In terms of technological developments, general milestones are displayed in Table 12.3.

FORMS OF RADIATION

Radiation is a form of energy. There are two basic types of radiation. One type is particulate radiation, which involves tiny fast-moving particles that have both energy and mass. Particulate radiation is primarily produced by disintegration of an unstable atom and includes alpha and beta particles. Alpha particles are high energy, large subatomic structures of protons and neutrons. They can travel only a short distance and can be stopped by a piece of paper or skin. Beta particles are fast moving electrons. They are a fraction of the size of alpha particles, but can travel farther and are more penetrating.

Rutherford, an English scientist, discovered alpha particles in 1899 while working with uranium. An alpha particle is identical to a helium nucleus having two protons and two neutrons. It is a relatively heavy, high-energy particle, with a positive charge of +2 from its two protons. Alpha particles have a velocity in air of approximately one-twentieth the speed of light, depending upon the individual particle's energy. Alpha particles don't get very far in the environment. They lose energy rapidly in air, usually expending it within a few centimeters. Because alpha particles are not radioactive, once they have lost their energy, they pick up free electrons and become helium (USEPA). Beta particles are subatomic particles ejected from the nucleus of some radioactive atoms. They are equivalent to electrons. The difference is that beta particles originate in

TABLE 12.1. Chronology of Food Irradiation.

1905	Scientists receive patents for a food preservative process that uses ionizing radiation to kill bacteria in food.
1921	U.S. patent is granted for a process to kill Trichinella spiralis in meat by using X-ray technology.
1953–1980	The U.S. government forms the National Food Irradiation Program. Under this program, the U.S. Army and the Atomic Energy Commission sponsor many research projects on food irradiation.
1958	The Food, Drug, and Cosmetic Act is amended and defines sources of radiation intended for use in processing food as a new food additive. Act administered by FDA.
1963	FDA approves irradiation to control insects in wheat and flour.
1964	FDA approves irradiation to inhibit sprouting in white potatoes.
1964–1968	The U.S. Army and the Atomic Energy Commission petition FDA to approve the irradiation of several packaging materials.
1966	The U.S. Army and USDA petition FDA to approve the irradiation of ham.
1971	FDA approves the irradiation of several packaging materials based in the 1964–68 petition by the U.S. Army and the Atomic Energy Commission.
1976	The U.S. Army contracts with commercial companies to study the wholesomeness of irradiated ham, pork and chicken.
1980	USDA inherits the U.S. Army's food irradiation program.
1985	FDA approves irradiation at specific doses to control Trichinella spiralis in pork.
1986	FDA approves irradiation at specific doses to delay maturation, inhibit growth and disinfect foods, including vegetables and spices. The Federal Meat Inspection Act is amended to permit gamma radiation to control Trichinella spiralis in fresh or previously frozen pork. Law is administered by USDA.
1990	FDA approves irradiation for poultry to control salmonella and other food-borne bacteria.
1992	USDA approves irradiation for poultry to control salmonella and other food-borne bacteria.
1997	FDA's regulations are amended to permit ionizing radiation to treat refrigerated or frozen uncooked meat, meat by products, and certain food products to control food-borne pathogens and to extend shelf life.
2000	USDA's regulations are amended to allow the irradiation of refrigerated and frozen uncooked meat, meat by products, and certain other meat food products to reduce the levels of food-borne pathogens and to extend shelf life. FDA's regulations are amended to permit the irradiation of fresh shell eggs to control salmonella.

Source: United States General Accounting Office (USEPA) (http://www.epa.gov/rpdweb00/sources/food_history.html).

the nucleus and electrons originate outside the nucleus. Henri Becquerel is credited with the discovery of beta particles. In 1900, he showed that beta particles were identical to electrons, which had just been discovered by Joseph John Thompson. Beta particles have an electrical charge of -1. The speed of individual beta particles depends on how much energy they have, and varies over a wide range. It is their excess energy, in the form of speed that causes harm to living cells. When transferred, this energy can break chemical bonds and form ions (USEPA).

The second basic type of radiation is electromagnetic radiation. This kind of radiation is pure energy with no mass and is like vibrating or pulsating waves of electrical and magnetic energy. Electromagnetic

TABLE 12.2. History of Food Irradiation (Wikipedia).

1895	Wilhelm Conrad Röntgen discovers X-rays ("bremsstrahlung", from German for radiation produced by deceleration)
1896	Antoine Henri Becquerel discovers natural radioactivity; Minck proposes the therapeutic use
1904	Samuel Prescott describes the bactericide effects at Massachusetts Institute of Technology (MIT)
1906	Appleby & Banks: UK patent to use radioactive isotopes to irradiate particulate food in a flowing bed
1918	Gillett: U.S. Patent to use X-rays for the preservation of food
1921	Schwartz describes the elimination of Trichinella from food
1930	Wuest: French patent on food irradiation
1943	MIT becomes active in the field of food preservation for the U.S. Army
1951	U.S. Atomic Energy Commission begins to co-ordinate national research activities
1958	World first commercial food irradiation (spices) at Stuttgart, Germany
1970	Establishment of the International Food Irradiation Project (IFIP), head quarters at the Federal Research Centre for Food Preservation, Karlsruhe, Germany
1980	FAO/IAEA/WHO Joint Expert Committee on Food Irradiation recommends the clearance generally up to 10 kGy "overall average dose"
1981/1983	End of IFIP after reaching its goals
1983	Codex Alimentarius General Standard for Irradiated Foods: any food at a maximum "overall average dose" of 10 kGy
1984	International Consultative Group on Food Irradiation (ICGFI) becomes the successor of IFIP
1997	FAO/IAEA/WHO Joint Study Group on High-Dose Irradiation recommends to lift any upper dose limit
2003	Codex Alimentarius General Standard for Irradiated Foods: no longer any upper dose limit
2004	ICGFI ends

Source: July 31, 2012 http://en.wikipedia.org/wiki/Food_irradiation

TABLE 12.3. Technological Developments in Radiation Processing.

1920–40	X-ray tube, medical applications, Therapeutic treatment of cancer
1940–50	Radiology equipment, medical research, diagnostic and treatment applications
1950–60	Cobalt-60, cesium-137, electron beam irradiators, medical and dental applications, sterilization of non-food items—surgical equipment and supplies and cosmetics
1960–70	Major food research applications, industrial irradiators, wider application in other areas—polymer cross linking, textiles, leather, tires
1970–90	Widespread applications, small and industrial scale equipment, regulatory clearances for food processing, controversial issues—risk/benefit analysis, consumer resistance/acceptance, focused market research
1990–2010	Sophisticated irradiation equipment—lab and industrial scale, wider clearances, wider acceptance, wider lobbying against, environmental and safety concerns

waves are produced by a vibrating electric charge and as such, they consist of both an electrical and a magnetic component. In addition to acting like waves, electromagnetic radiation acts like a stream of small "packets" of energy called photons. Another way that electromagnetic radiation has been described is in terms of a stream of photons. The massless photon particles travel in a wave-like pattern. Each photon contains a certain amount (or bundle) of energy, and all electromagnetic radiation consists of these photons. The only difference between the various types of electromagnetic radiation is the amount of energy found in the photons. Electromagnetic radiation travels in a straight line at the speed of light (3×10^8 m/s).

Physicists credit French physicist Henri Becquerel with discovering gamma radiation. In 1896, he discovered that uranium minerals could penetrate to a photographic plate through a heavy opaque paper. Roentgen recently had discovered x-rays, and Becquerel reasoned that uranium emitted an invisible light similar to x-rays. He called it "metallic phosphorescence." In reality, Becquerel had found gamma radiation being emitted by radium-226. Radium-226 is part of the uranium decay chain and commonly occurs with uranium. Gamma rays and x-rays, like visible, infrared and ultraviolet light, are part of the electromagnetic spectrum. While gamma rays and x-rays pose the same kind of hazard, they differ in their origin. Gamma rays originate in the nucleus. X-rays originate in the electron fields surrounding the nucleus or are machine-produced. Gamma radiation emission occurs when the nucleus of a radioactive atom has too much energy. It often follows the emission of a beta particle. Cesium-137 provides an example of radioactive decay by gamma radiation. When a neutron transforms to a proton and a beta particle, the additional proton changes the atom to barium-137. The nucleus ejects the beta particle. However, the nucleus still has too much energy and ejects a gamma photon (gamma radiation) to become more stable. Gamma emitting radionuclides are the most widely used radiation sources. The penetrating power of gamma photons has many applications. However, while gamma rays can penetrate many materials, they do not make them radioactive. The three radionuclides most useful are: Cobalt-60, Cesium-137, and Technetium-99m (USEPA).

Beta particle emission occurs when the ratio of neutrons to protons in the nucleus is too high. In this case, an excess neutron transforms into a proton and an electron. The proton stays in the nucleus and the electron is ejected energetically. This process decreases the number of neutrons by one and increases the number of protons by one. Since the number of protons in the nucleus of an atom determines the element, the conversion of a neutron to a proton actually changes the radionuclide to a different element. Often, gamma ray emission accompanies the emission of a beta particle. When the beta particle ejection doesn't rid the nucleus of the extra energy, the nucleus releases the remaining excess energy in the form of a gamma photon. Again, a photon is a discrete "packet" of pure electromagnetic energy. Photons have no mass and travel at the speed of light. The term "photon" was developed to describe energy when it acts like a particle (causing interactions at the molecular or atomic level), rather than a wave. Gamma and X-rays are high-energy photons.

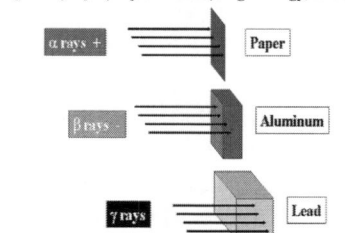

FIGURE 12.1. Forms of radiation from a radioactive source and their relative penetration power.

The penetration power of alpha, beta and gamma rays is illustrated in Figure 12.1. Alpha particles (positively charged particles) cannot penetrate most matter they encounter. Even a piece of paper or the dead outer layers of human skin is sufficient to stop alpha particles. The energy associated with beta rays (negatively charged steram of electrons) is much more and can be enhanced by increasing its velocity by acceleration under the influence of an electric field. Gamma rays are high energy photons and are true forms of electromagnetic radiations, a very high level of associated energy.

CHARACTERISTICS OF IONIZING RADIATIONS

The ionizing radiation being discussed is part of the electromagnetic spectrum (Figure 12.2) with radio waves on one end and the high energy X-rays and gamma-rays at the other. In the middle are the visible light rays, with infrared and ultraviolet rays on either side. Between the radio and infrared rays are microwaves. Since all these are part of the electromagnetic spectrum, they also will have several things in common. They are all waves with a characteristic wavelength, frequency and a certain amount of associated energy. The longer the wavelength, the smaller is the associated energy.

Radio and telecommunication waves with considerably long wavelengths of 30 cm to 3 km have very limited energy associated with them. Microwaves, although relatively low-energy waves, can cause molecular vibrations in materials like food that contain moisture or fat and can result in very rapid heating. Higher-frequency ultraviolet radiation begins to have enough energy to break chemical bonds. The X-rays and gamma rays, which are at the upper end of magnetic radiation, have very short wavelengths, high frequencies and very high associated energy levels. Radiation in this range has extremely high energy. When made to bombard materials, such waves can knock off an electron from an atom or molecule causing ionization. For this reason, these waves often are called ionizing radiations. When food irradiation is discussed, it is mainly with respect to ionizing radiations.

Ionization is the process in which a charged portion of a molecule (usually an electron) is given enough energy to break away from the atom. This process results in the formation of two charged particles or ions: the molecule with a net positive charge and the free electron with a negative charge. Each ionization releases approximately 33 electron volts (eV) of energy, and the material surrounding the atom

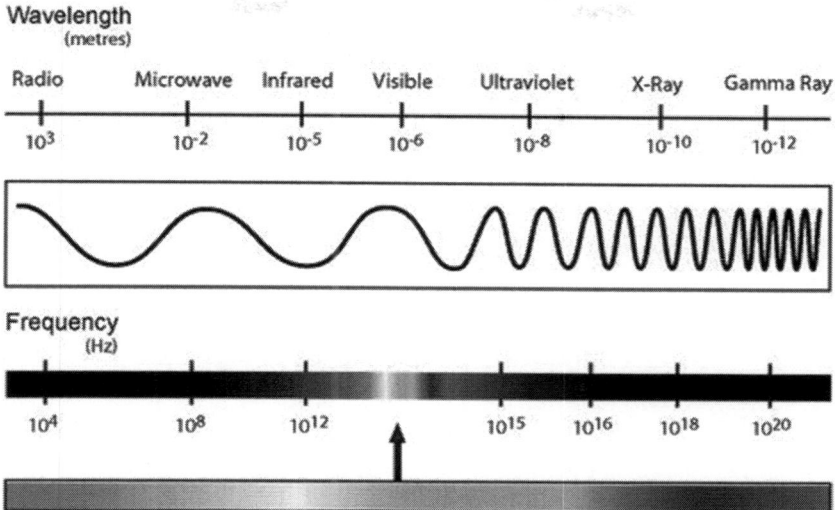

FIGURE 12.2. Electromagnietic spectrum.

absorbs the energy. Compared to other types of radiation that may be absorbed, ionizing radiation deposits a large amount of energy into a small area. In fact, the 33 eV from one ionization is more than enough to disrupt the chemical bond between two carbon atoms. All ionizing radiations are capable, directly or indirectly, of removing electrons from most molecules.

Relationships between frequency (ν), wavelength (λ) and energy associated with electromagnetic radiations are given below:

$$\nu = C/\lambda \tag{1}$$

where C = Velocity of light = 3×10^{10} cm/s

For γ rays, $\lambda = 3 \times 10^{-11}$ cm. Therefore,

$$\nu = \frac{3 \times 10^{10} \text{ cm/s}}{3 \times 10^{-11} \text{ cm}} = 10^{21} \text{ s}^{-1} \tag{2}$$

$$E = \eta\nu$$

where η = Plank's Constant = 4×10^{-15} eV.s

Therefore, for γ rays, E = $(4 \times 10^{-15}$ eV.s$) \times (10^{21}$ s$^{-1})$ or 4×10^{6} eV or 4 MeV.

SOURCES OF IONIZING RADIATIONS

The main source of gamma radiation is cobalt-60, a radioactive isotope produced from cobalt-59. A second source is cesium-137, which is a spent fuel of a nuclear reactor and also produces gamma rays. A third source is β-rays, which are a stream of electrons. Since the associated energy levels of these rays are too low to be of any practical value in terms of radiation preservation, they need to be accelerated (in cyclotrons, linear accelerators, etc.) to make them acquire the required energy. Careful precautions must be taken to ensure that all electrons have the exact amount of energy. If the acquired energy is too high, induced radioactivity in foods could occur upon irradiation.

Cobalt-60

George Brandt, a Swedish scientist, demonstrated in 1935 that a blue color common in colored glass was caused by a new element, cobalt. Previously, it was thought that bismuth, which occurs in nature with cobalt, was the cause. Radioactive cobalt-60 was discovered by Glenn Seaborg and John Livingood at the University of California Berkeley in the late 1930s (USEPA).

Non-radioactive cobalt occurs naturally in various minerals, and has been used for thousands of years to impart blue color to ceramic and glass. Cobalt-60 is produced by bombarding a sample of cobalt-59 with an excess of neutrons in a nuclear reactor. The cobalt-59 atoms absorb some of the neutrons and increase their atomic weight by one to produce the radioisotope cobalt-60. This process is known as activation. As a material rids itself of atomic particles to return to a balanced state, energy is released in the form of gamma rays and sometimes alpha or beta particles (USEPA).

In Canada, cobalt-60 is produced by AECL (MDS Nordion) in CANDU® (Canada Deuterium-Uranium) reactors (Figure 12.3). AECL's CANDU Reactor is heavy-water moderated and fueled with natural uranium. For cobalt-60 production, the reactor's full complement of stainless steel adjusters is replaced with neutronically equivalent cobalt-59 adjusters. Only the centre rod and stainless steel cable lengths are special for each adjuster application, as determined by the number of bundles required due to its location in the reactor. Each cobalt adjuster rod assembly consists of a number of bundles, each bundle containing up to 6 cobalt "pencils." During the enrichment the cobalt-59 bundles

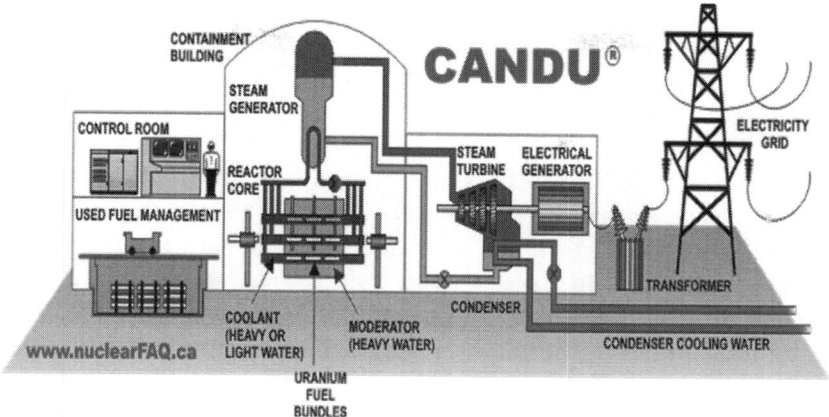

FIGURE 12.3. Canada Deuterium-Uranium reactor (MDS Nordion).

are continuously exposed (up to 2 years) to the moderated nuclear reaction and in the process acquire the extra neutron that transforms them to cobalt-60.

Cesium-137

Cesium (chemical symbol Cs) is a metal that may be stable (nonradioactive) or unstable (radioactive). The most common radioactive form of cesium is cesium-137. In 1860, Gustav Kirchhoff and Robert Bunsen discovered nonradioactive cesium in mineral water in Germany. Radioactive cesium-137, and many other radionuclides that are used in nuclear medicine, were discovered in the late 1930s by Glenn T. Seaborg and his coworker, Margaret Melhase. Nonradioactive cesium occurs naturally in various minerals. Radioactive cesium-137 is produced when uranium and plutonium absorb neutrons and undergo fission. Examples of the uses of this process are nuclear reactors and nuclear weapons. The splitting of uranium and plutonium in fission creates numerous fission products. Cesium-137 is one of the more well-known fission products.

Downgrading of Cobalt-60 and Cesium-137

The transformation from non-radioactive cobalt-59 to radioactive cobalt-60, and its subsequent degradation to nickel-60 (and cesium-137 degradation to barium-197), are shown in Figure 12.4. This process is known as radioactive decay, in which the radioactive element progressively loses its radioactivity to turn to its ground state.

Carrier Irradiators

Radiation processing of food is carried out inside an irradiation chamber shielded by 1.5–1.8 m thick concrete walls. Food, either prepacked or in-bulk in suitable containers, is sent into the irradiation chamber with the help of a conveyor. The conveyor goes through a concrete wall labyrinth, which prevents radiation from reaching the work area and the operator room. When the facility is not in use, the radiation source is stored 6 m deep under water. The water shield does not allow radiation to escape into the irradiation chamber, thus permitting free access to personnel to carry out maintenance. For treating food, the source is brought to the irradiation position above the water level after activation of all safety devices. The goods in aluminium carriers or tote boxes are mechanically positioned around the source rack and are turned round their own axis, so that contents are irradiated on both the sides. The absorbed dose is determined by the residence time of the carrier or tote box in irradiation position. Absorbed doses are checked by placing dosimeters at various positions in the tote box or carrier (MDS Nordion). Some details of the Carrier Irradiation System (MDS Nordion) are shown in Figure 12.5.

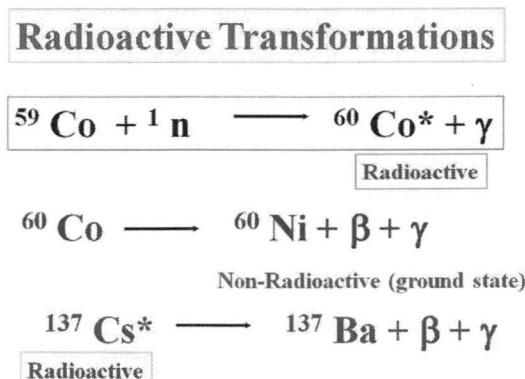

FIGURE 12.4. Radioactive transformations involving Cobalt and Cesium.

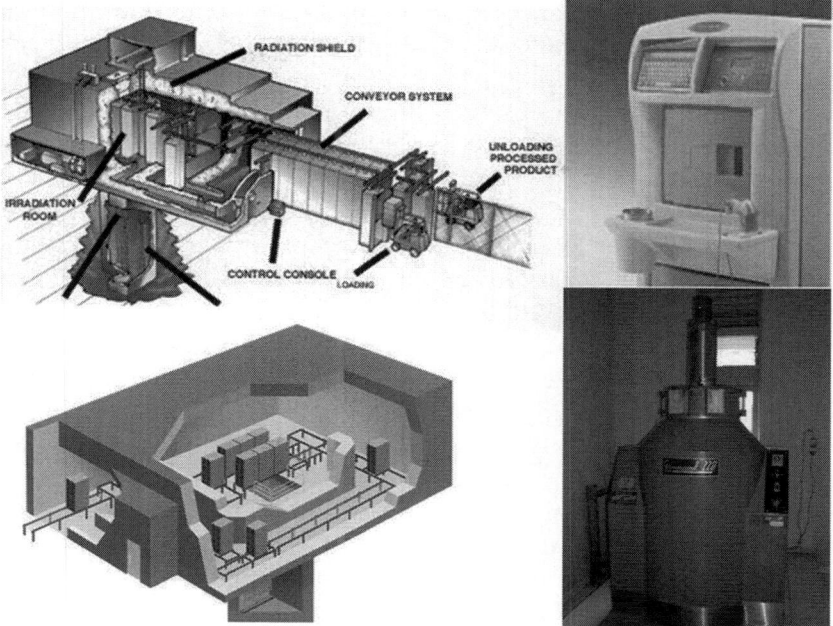

FIGURE 12.5. Industrial and laboratory irradiators based on Cobolt-60 (MDS Nordion).

Electron Beam (β-Rays)

Electron beam facilities generate e-beams with an electron beam linear accelerator. The electrons are concentrated and accelerated to 99% of the speed of light and to energies of up to 10 MeV. The electron beam works on the same principle as a television tube. Instead of being widely dispersed and hitting a phosphorescent screen at low energy levels, the electrons are concentrated. A conveyer moves the product to be irradiated under the electron beam at a predetermined speed to obtain the desired dosage.

Electron accelerators consist of a source (of electrons), an evacuated accelerating chamber and a system for extraction from the vacuum and distribution over the product surface. Most obtain their electrons from a heated filament source (similar to that of a TV tube) called the electron gun. The energy of these electrons is then increased in one or more stages as they pass through a vacuum with an applied electric field. There are numerous ways to generate this electric field. DC accelerators generate and maintain the full accelerating voltage between just two electrodes. As the voltage is raised to millions of volts, electrical insulation becomes a major engineering problem. Even at low powers, these accelerators can have dimensions exceeding 15m (45 feet). For systems with both the energy and the power needed for industrial irradiation, the most common commercially available models are the Dynamitron® and Insulated Core Transformer (ICT). The maximum voltage available is usually less than 5 MeV.

A second family of particle accelerators is based on radio frequency (RF) power technology. These high frequency waves generate very intense electrical and magnetic fields in suitably shaped conducting cavities. By matching the field oscillations with the injection of charged particles, the RF fields can drive the particles to high energies without having to create the full final potential at any one instant. They do not therefore require the large scale insulation of DC units and are more compact. They can easily reach energies of 10 MeV and currently are available in models reaching up to 200 kW of electron beam power. The RF accelerators that employ a linear series of RF cavities are called linacs. The assembly of copper cavities is referred to variously as a structure, a guide or a waveguide. The accelerating structure of a typical 10 MeV industrial linac is 2–4m long. The main power tube for linacs is the klystron. These tubes are expensive and have limited operating life. A more compact RF accelerator (Rhodotron®) has been developed. This device generates radial accelerating fields in a single cavity and uses an array of magnets to pass the beam through this cavity zone on repeated orbits. The accelerating structure for the higher-power models is a tank of about 1.5 m diameter systems. Differences in physical size between DC and RF accelerators are significant because DC accelerators require costly larger buildings and shields. Differences between the RF models are less important once the total cost of the shielded facility has been included (I-Ax Technologies Inc., Ottawa). Figure

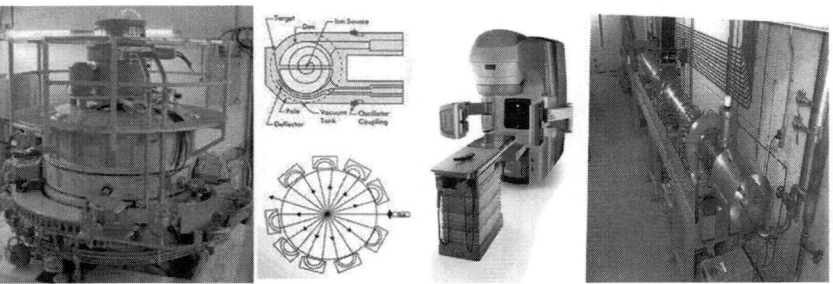

FIGURE 12.6. β ray accelerators and E-beam applicators.

12.6 shows commercial electron beam equipment which is becoming widely available.

Because e-beams are generated electrically, they offer certain advantages: (1) they can be turned on only as needed, (2) they do not require replenishment of the source as cobalt-60 does, (3) there is no radioactive waste. E-beam technology also have disadvantages: (1) shallow depth of penetration, (2) e-beams must be converted to x-rays to penetrate large items such as carcasses, (3) high electric power consumption, (4) complexity, and potentially high maintenance.

UNITS OF RADIATION

There are two types of units for measuring radiation: intensity and dose. The first one refers to the source and the second one to the product.

Intensity: The intensity or activity of a radioactive source is measured in number of disintegrations per second. The unit of intensity is curie, named in honor of Marie Curie. A curie (Ci) is defined as 37 billion disintegrations per second. The curie originally was a comparison of the activity of a sample to the activity of one gram of radium, which at the time was measured as 37 billion disintegrations per second. When more accurate techniques measured a slightly different activity for radium, the reference to radium was dropped.

Radiation is the energy that is released as high energy particles or rays, during radioactive decay. Radioactivity is the property of an atom that describes spontaneous changes in its nucleus that create a different nuclide. These changes usually happen as emissions of alpha or beta particles and often gamma rays. The rate of emission is referred to as a material's activity. Each occurrence of a nucleus emitting particles or energy is referred to as disintegration. The number of disintegrations per unit time is called the activity or intensity of a sample. Since each disintegration transforms the atom to a new nuclide, "transformation" often is substituted for "disintegration" when discussing about radioactive decay and activity.

The radioactive source is different from a conventional energy source since it continuously emits radiation. As it does, it decays itself and the process is recognized as radioactive decay. The radioactive decay is assumed to follow a first order kinetics and hence the changes in intensity (represented by either I or N) with respect to time can be represented by

$$-\frac{dN}{dt} \alpha N \qquad (3)$$

If N_o is the initial intensity, N is the intensity at time t, 'α' is the decay constant, then:

$$N = N_o e^{-at} \qquad (4)$$

The following graph (Figure 12.7) illustrates the first order or semi-logarithmic change in intensity N with time t.

Half-Life: The rate of radioactive decay is characteristic of each radionuclide. This generally is expressed in terms of a half-life. It is the time required for the disintegration of one-half of the radioactive atoms that are present when measurement starts. It does not represent a fixed number of atoms that disintegrate, but a fraction.

Half-life (λ) of a radioactive isotope is defined as the time interval between which the intensity of a radioactive source is reduced by one half. So by definition, the activity N_o is reduced to $N_o/2$ after a duration t equal to half-life (λ). S_o:

$$\frac{N_o}{2} = N_o e^{-a\lambda} \qquad e^{-a\lambda} = \frac{1}{2} \qquad (5)$$

Therefore:

$$a = \frac{0.693}{\lambda} \qquad (6)$$

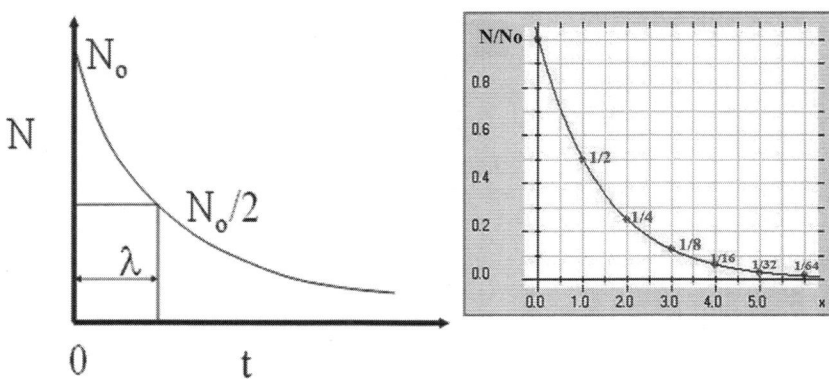

FIGURE 12.7. Radioactive decay of a radioisotope.

Half-life of Selected Radio Isotopes

The half-life of different radioactive sources varies. Cobalt-60 has a half-life of 5.27 years, while cesium-137 has a half-life of 30.2 years. From an operational point of view, longer half-lifes are better because they permit longer durations between adding more source to upgrade the activities, generally done after each half-life cycle. The intensity of radioactive source also varies; cobalt-60 has a higher intensity than cesium-137. Radium, a weak radioactive source has a half-life of 1600 years.

Intensity of Radioactive Source Decreases with Time

By combining two earlier equations, a useful equation that can be used to compute the existing activity (N) of a radioactive isotope with an initial activity of N_o and half-life λ as a function of time, t, can be represented as:

$$N = N_o e^{\frac{-0.693t}{\lambda}} \tag{7}$$

The activity or intensity when reduced by 50% after the passage of time is equal to the half-life of the radio isotope. It should not be assumed that after two half-life cycles, no activity is left. While the activity reduces to 50% after the first half cycle, it reduces to half of 50% after the second which means the residual activity will be 25%, and after third half-life, this reduces to 12.5% and so on. The changes in the activity and half-life pattern are shown in Figure 12.7.

Intensity Decreases with Distance from the Radioactive Source

As with time, radioactive intensity decreases semi-logarithmically with distance as the point of interest moves away from the radiation source. An equation similar to Equation (7) can be used to calculate the intensity of radiation (I) at a distance x from the source with Io representing the intial intensity.

$$I = I_o e^{(-\mu x)} \qquad a = \frac{0.693}{\lambda} \tag{8}$$

μ is defined as the absorption coefficient and λ can again be defined as half-distance, meaning the distance from the source that would reduce the intensity by 50%. The absorption coefficient for air is 0.00005 cm^{-1}; water 0.067 cm^{-1}.

Intensity Decreases within an Absorbing Body

The logic of Equation (8) can be extended to a body of material absorbing the radiation. The intensity will be maximum at the surface and will decrease exponentially with thickness as it penetrates into the body. Representing I as the intensity at distance x from the surface exposed to the radiation intensity (I_o), the same equation can be used to compute I. μ will be the radiation absorption coefficient of the object and l is the half-thickness, meaning thickness that will reduce the intensity by one half. The absorption coefficient of aluminum is 0.16 cm^{-1}, iron 0.44 cm^{-1} and lead 0.77 cm^{-1}; thus, lead has a good radiation absorbing capacity and is often used as the carrier material for transportation of radioactive materials and spent fuels.

RADIATION DOSE

The treatment received by the food product is characterized by the radiation dose; it is the quantity of energy absorbed by the food while it is exposed to the radiation field. It is expressed either in Rad or Gray. One Rad is the amount of radiation that results in the absorption of 10^{-5} J/g or 10^{-2} J/kg of radiation energy at the point of interest. The international unit of measurement is the Gray (Gy). One Gray represents one joule of energy absorbed per kilogram of irradiated product. One Gy is equivalent to 100 Rad. The amout of radiation given is dependent on the intensity of the source and the time of exposure.

The intensity of source, which normally is expressed in Curie (Ci), also is expressed in terms of a dose rate for practical purposes. The dose achieved is dependent on the time of exposure and the location of the product relative to the source, decreasing exponentially with the distance from source. The amount of energy absorbed by the food also will depend upon the mass, bulk density, and thickness of the food. For each kind of food, a specific dose has to be delivered to achieve a desired result. If the dose is less than appropriate, the intended preservation effect may not be achieved. If the dose is excessive, the food may be damaged and unacceptable for consumption. A dose rate is established for practical purposes based on the configuration of the set up. This is related to the amount of dose absorbed by a test sample placed in the chamber under fully loaded conditions. The dose rate decreases at the same rate as the intensity and hence the earlier equation with intensity also can be used with dose rate:

$$\text{Dose delivered} = \text{Dose rate} \times \text{time}$$

The equation used [Equation (8)] for intensity decline within an absorbing body also can be used to monitor changes in the resulting doses.

Example 1: The following data are for a Gamma cell 220 manufactured on January 1, 2000.

Co-60 activity:	25000 Ci
Full chamber dose rate:	1.26 Mrad/h
Chamber size:	6" diam × 8" length
Chamber load:	3.5 kg
Half-life of Co-60:	5.27 years

1. What would the residual Co-60 activity on January 1, 2012?
2. Calculate the irradiation time required to give a dose of 500 krad when installed.
3. Calculate the irradiation time required to give a dose of 500 krad on January 1, 2012.

Solution:

$$N = N_o e^{\frac{-0.693 t}{\lambda}}$$

1. We know that

   $N_o = 25000$ Ci
   $t =$ From January 1, 2000 to January 1, 2012 $= 12$ years
   $\lambda = 5.27$ years
   N can be calculated to be 5160 Ci

2. We can calculate the irradiation time by dividing dose by dose rate

   $$\text{Irradiation time} = \frac{\text{dose}}{\text{dose rate}} = \frac{500 \text{ krad}}{1260 \text{ krad/h}} = 0.4 \text{ h} = 23.8 \text{ min}$$

3. Irradiation time for a dose of 500 krad on January 1, 2012.

Since the intensity decreases from 25000 to 5160 Ci, the dose rate decreases by the same margin. So the new dose rate will be: 1260 × (5/160/25000) krad/h or 10^3 krad/h and therefore the new time 500 krad/10^3 (krad/h) or 4.8 h, which is more than 12 times the original irradiation time.

Example 2: A cobalt 60-source rack is immersed in a well 10 m deep. It has an initial activity of 100 MegaCi capable of giving a dose of 500 MGy/s if exposed next to the source. The half thickness of water for absorption of radiation is 10 cm.

1. What will be the dose absorbed by a person working outside the well for 12 h for a major repair to some equipment ?
2. After how long will the radiation absorbed by a person reach a level of 1 mGy?

Solution: The problem is represented by Figure 12.8.

1. The intensity change with respect to absorption within a medium is

$$I = I_o e^{(-\mu x)}$$

where x is thickness, μ is absorption coefficient and I and I_o are the absorption at the source and at a distance x from the source.

We also know: $\mu = 0.693/\lambda$, and the half-thickness λ is 10 cm or 0.1 m

$$\mu = 0.693/0.1 = 6.93 \text{ m}^{-1}$$

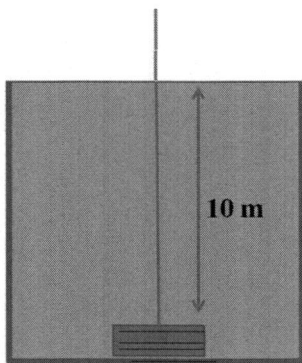

FIGURE 12.8. Source rack in a water pool.

Substituting for m and I_o and x, we find the reduced dose at the surface as

$$I = (500 \text{ MGy/s}) \, e^{-(6.93 \times 10)}$$

$$I = 4.0 \times 10^{-28} \text{ MGy/s}$$

2. After how long will the absorbed dose reach 1 mGy level (really small level)?

Dose rate: 4.0×10^{-28} MGy/s or 4.0×10^{-19} mGy/s since 1 MGy = 10^9 mGy

Dose = 1 mGy

Time = dose/dose rate = 1 mGy/[4.0×10^{-19} mGy/s] or 2.5×10^{18} s or 79 trillion years

Example 3: If the irradiator in example 2 with 100 MegaCi source is exposed to air by accident, what would be the radiation intensity a residential community 2.5 kilometer away is exposed to? The abosorption coefficient of radiation by air is 0.00005 cm^{-1}

Solution: This is similar to problem 2.

$$I = I_o e^{(-\mu x)}$$

I_o = 100 MegaCi, μ = 0.00005 cm^{-1}, x = 5.0 km or 1000 m or 1000000 cm

I = 100 MegaCi × exp (–0.00005 cm^{-1} × 500000 cm) = 1.38E-09 MegaCi

Assuming as in the previous example, the full intensity to give a dose rate of 500 MGy/s, the residual activity of 1.38E-09 MegaCi will yield a dose rate of 6.94E-09 MGy/s which accumulates to 0.6 kGy per day, almost like the low dose irradiation given for fruits and vegetables.

Labeling

The Food and Drug Administration (FDA) has established regulations for labeling of irradiated foods. Labels must contain the words "Treated with Radiation" or "Treated by Irradiation" and display the irradiation logo, the Radura (Figure 12.9). The petals represent the food, the central circle the radiation source, and the broken circle illustrates the rays from the energy source. The FDA requires labeling of pack-

FIGURE 12.9. Internationally accepted symbols for irradiated foods.

aged, irradiated food sold at retail stores. Irradiated whole foods sold in bulk, such as fruits and vegetables, also must display the label. No label is required for food products that contain irradiated ingredients, such as spices, as long as the entire product has not been irradiated. Irradiated foods sold at the wholesale level also must be labeled. However, both the shipping container and the invoice or bill of lading must display the statement, "Do not irradiate again." The FDA has not evaluated products that have been irradiated more than once. The FDA does not require labeling of irradiated food served in restaurants.

Classification of Doses

Radiation doses are classified into low, medium and high. For fruits and vegetables, low doses of up to 1 kGy are used: to inhibit sprouting of certain crops such as potatoes and onions; to disinfect fruits and vegetables from insects and parasites and to delay physiological processes (e.g., ripening) of fruits. A delay in postharvest ripening can occur only in a climacteric fruit, which ripens normally after harvest. To eliminate spoilage microorganisms and to extend the shelflife of fruits and vegetables, medium doses of 1 to 10 kGy are necessary. Fungi and bacteria are known to have serious pathological effects on fruits. Senescence and physiological breakdown occur both for climacteric and nonclimacteric fruits as they age in postharvest storage. High doses of 10 to 50 kGy are used to decontaminate herbs, spices and food ingredients. At dose levels where the growth of spoilage microorganisms such as fungi on a fruit can be controlled, a number of problems may be encountered because the dose levels used are always higher than those used for delaying ripening of fruits. The problems usually are related to the physiological response of the commodity being irradiated. Thus, an effective treatment is dependent on the relative sensitivity of the host (e.g. fruit) and the spoilage microorganism (e.g., fungi). The host must be able to tolerate the treatment without any apparent injury or other undesirable side effects. It is clear that ionizing energy has potential applications to fruit commodities, but it has limitations as well. In most cases for fruits, doses will be determined by the fruit species rather than by the pathogen.

How Hot is the Irradiation Process?

Conventional sterilization requires up to 20 min heating at 121°C using heat requiring pressurized steam heating conditions or hours of dry heating at temperatures in excess of 140°C. However, irradiation sterilization requires doses in the range of 20 kGy (commonly used for sterilization of spices). How hot is this process?

Since 1 Gy = 1 J/kg, an absorbed dose of 20 kGy means an absorbed energy of 20 kJ/kg by the product. This added energy will raise the temperature of the product. In a conventional sense, the3 energy is equivalent to the multiplication product of the treated sample mass, its heat capacity and temperature rise ($Q = mCp\Delta T$). Hence the T can be obtained by dividing the energy by-product of heat capacity and mass. Approximating the heat capacity as similar to that of water (4.2 kJ/kg°C), the temperature increase can be obtained as 20/4.2 or approximately 5°C. Hence, the sterilization process would only increase the product temperature by 5°C. That is the reason for calling the irradiaton sterilization process "cold sterilization."

Mechanisms of Action of Irradiation

The mechanism of action for irradiation has been attributed to free radical formation and a direct-hit target theory. The radiation energy (ionising radiation) penetrates the food and produces free radicals from the material through which it passes. Free radicals are highly reactive and very short-lived—so short-lived that they cannot even be detected in food almost immediately after it has been irradiated.

The target theory is based on the fact that gamma rays can hit and denature sensitive cell material, importantly DNA (deoxyribonucleic acid), the largest molecule in the nucleus and also RNA (ribonucleic acid). DNA consists of a very long ladder twisted into a double helix. The backbone is composed of sugar and phosphate molecules while the rings of the ladder are comprised of four nucleotide bases (cytosine, thymine, adenine and guanine), which are joined weakly in the middle by hydrogen bonds. Disruption of these weak hydrogen bonds prevents replication and causes cell death while exerting minimal effects on non-living tissue. Living organisms deprived of intact DNA or RNA will cease to function. Hence ionising radiation can slow down cell-based processes, such as early ripening in fruit. Likewise, it is effective against insects and molds.

Irradiation Application Areas

A number of applications for irradiation have been identified as improving safety and reducing food spoilage. The following are a few:

1. Low-dose (less than 1 kGy) irradiation is adequate for insect control.
2. Medium-dose irradiation (up to 3–7 kGy) for salmonella, E. coli and other food poisoning/pathogenic bacteria.
3. High-dose irradiation (10–30 kGy) for spices and herbs for disinfestation and for containing food poisoning and pathogenic microorganisms.
4. Low-dose irradiation (~1–2 kGy) for shelf-life extension and spoilage control.
5. Low-dose irradiation (~1 kGy) for sprout inhibition.
6. Low-dose irradiation for quarantine treatment for insect decontamination.
7. High-dose irradiation (20–45 kGy) for commercial sterilization.

Irradiation also has been used in many non-food applications like:

    a. Sterilization of surgical supplies and instruments.
    b. Sterilization of refrigerated seawater for fish storage.
    c. Polymer cross linking—textile, leather and tires.
    d. Sterilization of industrial waste.
    e. Sterilization of pharmaceutical base products.

FOOD IRRADIATION WORLDWIDE

More than 90,000 tonnes of dried herbs, spices and vegetable seasonings were irradiated in some 20 countries in 2000, with around half of this quantity being irradiated in the USA. Food irradiation is classified as a food additive in USA legislation. Since 2002, beef also has been irradiated in the USA and sold to a growing market (IFT 2004, IFST 2006).

One major E-Beam facility in the USA overestimated the expected uptake of irradiated beef for the School Lunch Programme and went out of business in 2004. However, a Texas-based investment firm purchased the assets in June 2005 and in late December, the plant began processing about 40,000 pounds per day of animal feed for mills in the US Midwest. Fermented pork sausages (Nam) usually consumed raw in Thailand have been irradiated since 1986 (Food Standards Agency 2004).

A survey of the extent to which foods are irradiated in the EU, carried out by the Commission of the European Communities, (EC, 2004) found:

- Belgium irradiated 6,613 tonnes (frozen frogs' legs, frozen seafood and spices/seasonings were the principal products).
- Germany irradiated 795.3 tonnes (dried aromatic herbs and spices and herbal tea—exports to Poland comprised the majority of products).
- France irradiated 5,129 tonnes (mechanically recovered chicken meat, spices and frozen frogs' legs were the principal products). The Netherlands irradiated 7,114.4 tonnes (dehydrated vegetables, spices and herbs, frozen poultry and foods intended for export to third countries comprised the majority).
- No food was irradiated in the UK. In the UK, very few, if any, irradiated food products are on retail sale. A survey was carried out in 1996 and repeated in 2002 to investigate whether irradiated food is on sale in the UK but not labeled as such (Food Standards Agency, 2002). In this survey, 543 samples without declared irradiated ingredients covering three food categories (203 herbs and spices, 138 dietary supplements and 202 prawns and shrimps) were analysed. These three food categories were selected because a number of reports had claimed that these products were likely to have been irradiated and unlabeled. One of the herbs and spices (ground nutmeg), five prawns and shrimps and forty-four dietary supplements (42%) were found to be irradiated or to contain irradiated ingredients without appropriate labeling. The analytical methods, PSL (photostimulated luminescence) used as a screening procedure while TL (thermoluminescence) was used to confirm those samples, which gave

"positive" or "intermediate" (suspected irradiation) when analyzed by PSL. Comments were elicited from all the retailers or suppliers of the offending products. These varied from a declaration that the company does not knowingly sell irradiated food to queries over the accuracy of the analytical results or that an excipient ingredient (talc) may have been responsible for the false positive.

The Food (Control of Irradiation) Regulations came into force in 1991 and were amended in 2001. On 15 June 2004, the UK Food Standards Agency (FSA) issued an alert [Food Alert For Information (FAFI)] on noodle-based snacks due to the undeclared presence of irradiated ingredients contained in the vegetable seasoning mix of the dried spicy soup (IFST).

Radurization, Radicidation and Radappartization

Radurization: This refers to a low-dose irradiation treatment. Radurization is pasteurization by radiation. It is primarily used to treat foods that have high moisture content and a high pH. The microbes targeted are mainly spoilage organisms. Meat and fish are the foods for which this process is mainly used. In low-moisture acidic foods, yeasts and molds can be denatured. The treatment dose for radurization is approximately 1 kGy. When used for fresh fruits and vegetables, this level may result in shelf-life extension by slowing down various physiological processes. It is also used for sprout inhibition, for infestation control and related applications. The treated product must be refrigerated to further control microbial, enzymatic and physiological activities.

Radicidation: This makes use of medium-dose irradiation. The process of radicidation is used to eliminate pathogens and insect pests. This process kills vegetative cells only, meaning that it will not kill spores. Also, certain radiation-resistant vegetative cells can survive, including some strains of the bacterium Salmonella, which have been found to be radiation-resistant. The dose for radicidation ranges from 2.5 to 5.0 kGy. Here also, the product needs to be refrigerated post-treatment.

Radappertization: This makes use of high-dose irradiation. Radappertization involves treating the product to levels of radiation of approximately 30 kGy. This high level of radiation also destroys spores of microorganisms such as *Clostridium botulinum*. Such levels generally are deemed sufficient for clinical sterility, but not usually employed on food items. This will result in a shelf-stable product if the packaging provides an inert atmosphere and constitutes a barrier of gas, water vapour and microorganisms.

Decimal Reduction Value (D value)

Decimal reduction value (or D value) is a term traditionally used in thermal processing applications as an indicator of the thermal destruction rate. At any given temperature, it is the time required to reduce the surviving microbial population by one decimal, meaning it results in 90% destruction of the microbial population. It also is referred to as a logarithmic reduction.

The D-value with respect to irradiation indicates the killing power of radiation for a particular microbe or activity. One D-value is the amount of irradiation dose needed to kill 90% of the organisms. The semi-logarithmic survival plot of a microorganism subjected to irradiation can be represented by Figure 12.10, which is a plot of residual concentration of microorganism vs. dose. The D value can be represented as a slope index indicating the dose range required for the curve to pass through one log cycle. Mathematically, it is the negative reciprocal slope of the irradiation dose or the survivor curve.

For example, *E-coli* has a D-value of 0.3 kGy. *Clostridium botuli-*

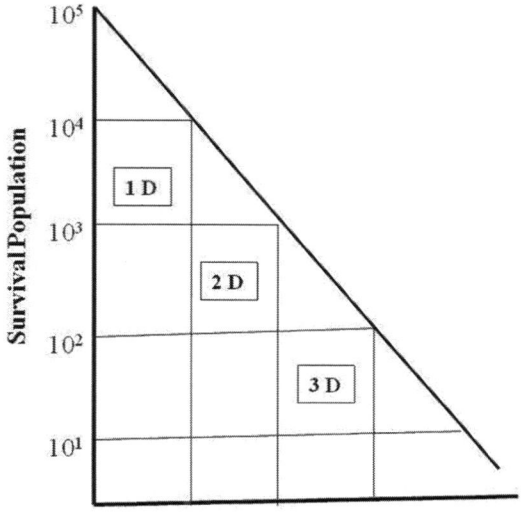

FIGURE 12.10. Irradiation dose survivor curve.

num (Type A) has a D value of 3.0–3.7 kGy. Thermal process experience recommends a 12 D process based on this strain for sterilization. Hence, to achieve the same thorough irradiation, an irradiation dose up to 35–45 kGy may be required.

Toxicological Safety of Irradiated Foods

An evaluation of toxicological food safety concerns arising from food irradiation has been published (IFT 2004, IFST 2006). Foods sterilized by high-dose irradiation (>25 kGy)—cold sterilisation as opposed to thermal sterilization (canning)—have been consumed by astronauts in the NASA space shuttle programme because of their superior quality and variety, compared to foods treated by other preservation technologies. There is a small but increasing demand for sterile products for immunocompromised patients as well as for niche markets, such as the military, campers or disaster victims where a long shelf life at ambient temperatures is required. High-dose sterile foods may be prepared under medical supervision for immunocompromised patients without labelling. The current upper limit of 10 kGy is insufficient to achieve sterility. This led a Joint FAO/IAEA/WHO Study Group on High-Dose Irradiation to request the International Consultative Group on Food Irradiation, to petition the Codex Secretariat to remove the upper dose limit of 10 kGy by revising the General Standard. This recommendation was based on the usefulness of effectively eliminating the more resistant spores of proteolytic strains of *Clostridium botulinum* as well as all vegetative organisms while neither compromising nutritional values nor resulting in any toxicological hazard. The process of cold sterilization is analogous to canning (thermal sterilization), the products of which have been safely consumed for well over a century. The conclusion reached by the Joint Study Group was that food irradiated to any dose appropriate to achieve the intended technological objectives is both safe to consume and nutritionally adequate. They also advised that no upper dose limit need be imposed. The Joint Study Group concluded that appropriate steps need to be taken to establish the technological guidelines implied by these conclusions and then to communicate them through Codex Alimentarius standards in order to achieve global standardization. The revised Codex General Standard (1998) for Irradiated Foods now reads, "The maximum absorbed dose delivered to a food shall not exceed 10 kGy, except when necessary to achieve a legitimate technological process."

Advantgages of Irradiation Processing

The following advantages have been ascribed to the irradiation process:

- Low energy consumption
- Environmental cleanliness
- Noise control
- Potential to reduce the use of chemical agents like fungicides, pesticides, fumigants, etc.
- Control of pathogenic and food poisoning microorganisms and reduce outbreaks
- High reliability
- Low overall cost

Radiation's Potential Disadvantages

1. *Nutritional quality loss:* Irradiation is very effective against living organisms that contain DNA and/or RNA, but does not cause any significant loss of macronutrients. Proteins, fats and carbohydrates undergo only small change in nutritional value during irradiation even with doses over 10 kGy, though there may be sensory changes. Similarly, the essential amino acids, essential fatty acids, minerals and trace elements also are unaffected. There can be decreases in certain vitamins (particularly thiamine) but these are of the same order of magnitude as occurs in other manufacturing processes, such as drying or canning (thermal sterilization).
2. *Mask spoilage in food:* There has been some concern that irradiation could be used as a means to mask spoiled foods. Although food irradiation kills microorganisms in foods that already are spoiled, it cannot suppress odors or other signs of spoilage, and thus cannot be used as a means to "hide" or "cover up" spoiled food. Many bacteria and other microorganisms that produce bad odors or discoloration will remain as a warning sign to consumers that food is spoiled, even after the food has been irradiated.
3. *Discourage strict adherence to good manufacturing practices:* As with any other food processing, food irradiation is not a substitute for safe food handling by processors, retailers and consumers. In addition, food irradiation should go hand-in-hand with modern Hazard Analysis and Critical Control Points (HACCP), a preventive food

safety management system that is mandated in meat, poultry and seafood processing plants in many countries. Consumers must also practice safe food handling techniques, whether a food is irradiated or not. It remains possible for bacteria to multiply in irradiated food if it has not been refrigerated properly or if care was not taken to avoid cross contamination with harmful bacteria from other sources.

4. *Radiological safety:* The safety of irradiation facilities is regulated by government bodies in every country. Nuclear regulatory boards oversee the safety of irradiation facilities and regulate their application. The incidents that have occurred in the past are documented by these agencies and thoroughly analyzed to determine root cause and improvement potential. Such improvements are then mandated to retrofit existing facilities and future design. Utmost care is taken not to expose the operators and the environment to radiation or radioactive contamination. Interlocks and safeguards are mandated to minimize this risk. Nevertheless there have been radiation-related deaths and injury among workers in such facilities, and these cause alarm. Nothing is 100% safe, and processes must be judged based on risk and benefit analysis. Lessons can be learned from nuclear disasters like those that have happened in USA, Russia and more recently in Japan.

5. *Induced radioactivity:* Induced radioactivity occurs when a previously stable material has been made radioactive by exposure to specific radiation. Mostly, radioactivity does not induce other material to become radioactive. Neutron activation is the main form of induced radioactivity, which happens when free neutrons are captured by nuclei. This happens during the process of creating radioactive materials and requires long-time exposure to high-intensity radiation. Irradiation of foods does not make them radioactive. The extent of induced radioactivity depends on the type and intensity of applied radiation, dose and the abundance of minerals. Generally, it has been accepted that cobalt-60 and electron beams below 10 MeV level do not have energy levels to induce radioactivity in irradiated foods. A less common form involves removing a neutron via photodisintegration. In this reaction, a high energy photon (γ-ray) strikes a nucleus with energy greater than the binding energy of the atom, releasing a neutron. This reaction has a minimum cutoff of around 10 MeV for most heavy nuclei. Many radionuclides do not produce gamma rays with energy high

enough to induce this reaction. The isotopes used in food irradiation (cobalt-60, cesium-137) both have energy peaks below this cutoff and thus will not induce radioactivity in the food. There may be some background radiation produced, which is small and disappears with time.

6. The conditions inside nuclear reactors can cause induced radioactivity. The components in those reactors may become highly radioactive from the radiation to which they are exposed. Induced radioactivity in such cases increases the amount of nuclear waste that must be properly disposed.
7. Multiple irradiations can be a problem and proper regulation should be in place to prevent it so that an overall applied dose does not exceed the specified limits.

Consumer Approval

The introduction of irradiated foods has many analogies with the introduction of pasteurization of milk over a century ago—one of the most significant advances ever made in food safety. The principal allegations advanced against the introduction of thermal pasteurization of milk and food irradiation (cold pasteurization) are very similar. Opponents of both thermal pasteurization of milk and cold pasteurization of foods by irradiation have claimed that: Nutritional value will be diminished, will be used to mask filthy products, legalizes the right to sell stale food, is unnecessary, is meddling with nature and will take the "life" out of the product.

Many surveys have been carried out (mostly in the USA) to assess consumer attitudes to food irradiation. Results have consistently shown that many consumers have misconceptions about the technology and believe that it makes food radioactive. When consumers are given information about the process and a chance to try irradiated products for themselves they are much more likely to accept the technology. Market trials also have met with success. Studies have indicated the irradiated products were in fact preferred over alternatives and consumers were willing to pay more for such foods.

Dosimetry

Measuring a dose is known as dosimetry. The three main purposes of dosimetry in food irradiation are: (1) to develop the proper dose requirement for the food commodity under research; (2) to assemble data for commissioning the food product through the regulatory agency; and (3) to establish the quality control procedure in the food production plant. It also is used to determine the configuration of the irradiation field after the installation of the irradiator or any changes that would occur in the irradiation facility.

In general, experimental dosimetry is preferred over calculation methods. Dosimeters are placed within the food product being irradiated to measure the distribution of the absorbed energy and to determine the maximum and the minimum dose absorbed by the food. The measurement of the absorbed dose must be accomplished as accurately as possible to establish correct procedures for food preservation and quality control in irradiation processing.

The choice of a dosimeter is mainly dependent upon: (1) its reliability for calibration and standardization; (2) its reproducibility at all specified dose levels; (3) its limited dependence on process conditions (i.e., radiation spectra, dose rate, etc.); (4) its possibility for correction of systematic errors caused by environmental factors (i.e., humidity, temperature, light effects, etc.); (5) its degree of equivalence to the food product; (6) its commercial availability at low cost and its ease of handling.

Primary or reference dosimeters (e.g., calorimeter) are used to calculate the amount of energy absorbed or the dose directly from the measurements. Secondary or routine dosimeters give measurements that vary with the amount of energy absorbed. The dose is estimated from a calibration curve in which measurements obtained from a secondary dosimeter are plotted against the measurements made under similar conditions by the primary dosimeter. Once secondary dosimeters are properly calibrated, they can be used to estimate the dose absorbed by the food.

Ceric-cerous is an aqueous chemical dosimeter. Ceric-cerous dosimeters are used mainly for research purposes. Upon irradiation, ceric ions are reduced to cerous ions. The readout system is simple and comprises an electrochemical cell and a digital multimeter. It is relatively inexpensive. The range of the dosimeter covers most food requirements (low dose: 0.6 to 10 kGy, high dose: 5 to 50 kGy). It usually is used in the development of the technology. *Fricke* is a dosimeter generally prepared in-house. The Fricke dosimeter is widely used for calibration and for comparison between laboratories. It is one of the most useful reference dosimeters for intercomparisons by national laboratories. Aqueous solutions of Fe^{+2} ions are converted into Fe^{+3} ions. An ultraviolet spectrophotometer is used as a readout system to determine the concentration of Fe^{+3} at 305nm. The range is 0.02 to 0.4kGy.

For routine dosimetry during processing of fruits, *FWT 60 radiochromic film* (Far West Technology) is the most commonly used dosimeter. Its ruggedness is excellent. Its size is small and it is relatively inexpensive. The dosimeter is usable throughout the anticipated dose range (0.1–100kGy) in food processing. Other commercial dosimeters available for routine measurement of the absorbed dose are the Harwell YR (Didcot, UK) and Farwest Optichromic (Goleta, CA).

DETECTION OF RADIATION

Detection methods are not required for clearance even though irradiation of various food products has been approved in more than 30 countries. Nevertheless, the development of methods to detect irradiation is important for a number of reasons, such as differences in national legislation, control of international trade, quality control and consumer reassurance. Several attempts have been made to detect irradiation on an experimental basis that has proven to be most difficult because the changes induced by irradiation are small and are not radiation specific. It is not likely that a universal method will be developed for all foods, but methods can be devised for specific groups. Willemot *et al.* (1996) reviewed in detail the detection methods for fruits. At present, the most promising irradiation detection methods are electron spin resonance (ESR) and thermoluminescence. Further studies in the development of instrumentation and standardization of methods eventually may lead to general practical applications.

REGULATORY APPROVAL

National governments enforce regulation of irradiated foods and irradiation facilities. Regulatory agencies determine which food may be treated by irradiation, under what conditions, and for what purpose. They also prescribe what type of information must be included on a label. Food irradiation plants are inspected like all food processing operations and have to comply with regulations by authorities responsible for the safety of the application of irradiation.

World

In 1983, the Codex Alimentarius Commission agreed that foods irradiated up to 10 kGy were safe and wholesome and therefore toxicological testing was no longer necessary. In 1988, the FAO, WHO, IAEA and the International Trade Centre-UNCTAD GATT jointly organized the international conference on the acceptance, control and trade in irradiated food in Geneva, Switzerland (IAEA, 1989). The use of food irradiation was endorsed by government designated experts from 57 countries. Since then, countries around the world have brought their regulations in line with the Codex General Standard of Irradiated Foods and have cleared many foods for irradiation. It is in South Africa that the greatest variety of fruits are processed annually with ionizing energy, including mangoes, papayas, bananas, litchis, tomatoes and strawberries. Although many attempts have been made to internationalize regulation of this technology, most countries continue to approve its use on a case-by-case basis. Today, at least 36 countries have collectively approved irradiation of more than 50 different foods.

USA

In 1986, the FDA issued a regulation (CFR title 21 part 179) that permits irradiation of fruits at a maximum dose up to 1 kGy for certain benefits such as insect disinfestation and ripening, growth and maturation inhibition. Not only specific applications to foods are specified in this regulation but also type and sources of radiation: gamma rays from cobalt-60 or cesium-137, accelerated electrons from a machine source not to exceed 10 MeV or x rays from a machine source not to exceed 5 MeV. Also, labeling is required. The internationally recognized logo (Figure 12.9) must appear on the package or the display as well as either of the following statement "treated with radiation" or "treated by irradiation". If irradiated fruits are shipped to a manufacturer for further processing, a label must be displayed "do not irradiate again". Anyone may irradiate fruits in compliance with this regulation without further permission from FDA. This regulation is a full authorization. There are no FDA licensing procedures for plant facilities. Plants must comply with current FDA regulations for good manufacturing practices for the production, handling and storage of foods (CFR title 21 part 110). FDA monitors food processing by periodic unannounced inspections of the facilities under its jurisdiction. Moreover, industrial activity must conform to laws designed to protect workers' safety and the environment. The use of a radioisotope such as cobalt-60 or cesium-137, as a source of radiation, requires a license issued by the US Nuclear Regulatory Commission (USNRC). In the case of electron accelerators and x rays machines, concerns regarding safety arise only when the machine is turned on. Thus, manufacturers of machine sources of radiation are not required to get a licence from the USNRC but they must submit a report to FDA's Center for Devices and Radiological Control for Health and Safety Act. While this imposes no requirement on food processors, a processor should ensure that the equipment has been reported by the manufacturer. It has to be pointed out that some states require registration or licensing for facilities using machine-generated radiation.

Table 12.4 lists the irradiation dose level permitted in USA and purposes of irradiation for various foods, while Table 12.5 displays the maximum dose allowable for different packaging materials.

TABLE 12.4. Foods Permitted to be Irradiated Under 21 CFR 179.26(b) as of July 2012.

Food	Purpose	Dose
Fresh, non-heated processed pork	Control of *Trichinella spiralis*	0.3 kGy min. to 1 kGy max.
Fresh foods	Growth and maturation inhibition	1 kGy max.
All foods	Arthropod disinfestation	1 kGy max.
Dry or dehydrated enzyme preparations	Microbial disinfection	10 kGy max.
Dry or dehydrated spices/seasonings	Microbial disinfection	30 kGy max.
Fresh or frozen, uncooked poultry products	Pathogen control	3 kGy max
Frozen packaged meats (solely NASA)	Sterilization	44 kGy min.
Refrigerated, uncooked meat products	Pathogen control	4.5 kGy max.
Frozen uncooked meat products	Pathogen control	7 kGy max.
Fresh shell eggs	Control of *Salmonella*	3.0 kGy max.
Seeds for sprouting	Control of microbial pathogens	8.0 kGy max.
Fresh or frozen molluscan shellfish	Control of *Vibrio* species and other foodborne pathogens	5.5 kGy max.
Fresh iceberg lettuce and fresh spinach	Control of food-borne pathogens and extension of shelf-life	4.0 kGy max.

TABLE 12.5. Pre-packaging Materials Permitted for Irradiated Foods (21 CFR 179.45).

Section	Packaging Materials	Max Dose [kGy]
179.45(b)	Nitrocellulose-coated cellophane	10
	Glassine paper	10
	Wax-coated paperboard	10
	Polyolefin film	10
	Kraft paper	0.5
	Polyethylene terephthalate film	10
	Polystyrene film	10
	Rubber hydrochloride film	10
	Vinylidene chloride-vinyl chloride copolymer film	10
	Nylon 11 [polyamide-11]	10
179.45(c)	Ethylene-vinyl acetate copolymer	30
179.45(d)	Vegetable parchment	60
	Polyethylene film	60
	Polyethylene terephthalate film	60
	Nylon 6 [polyamide-6]	60
	Vinyl chloride-vinyl acetate copolymer film	60

Canada

In Canada, products must be irradiated in facilities specially constructed for this purpose. These plants must possess a licence issued by the Atomic Energy Control Board (AECB), which inspects the facilities regularly to verify safety and compliance. Further, irradiated food products must be approved in advance by Health Canada. In the current legislation, a dose of 0.15 kGy may be used to inhibit germination of potatoes and onions.

Wheat, flour and whole wheat flour may be irradiated with a dose of 0.75 kGy to prevent insect infestation during storage. Whole or ground spices and dehydrated seasonings may be irradiated with a maximum dose of 10 kGy to reduce the initial microbial charge. Today, irradiation of fruits is not permitted by the law. To get permisson to irradiate fruits, processors must prepare a submission to Health Canada according to the pre-clearance requirements for proposed irradiated food. Details including the purpose of the proposed irradiation, the irradiation dose required, the chemical, physical, microbial and nutritional effects, the details of other processes to be applied before and after irradiation and recommended conditons of storage and shipment must be specified.

Then, the petition is evaluated in terms of whether it addresses and satisfies the pre-clearance requirements. If accepted, the petition will follow due regulatory process, and the fruit will be put on the clearance list for irradiation in the Canadian Food and Drug Regulations. If the irradiated food is sold in the marketplace, it also must comply with labeling requirements. Identification of an irradiated food ingredient, using the international symbol, is required if it makes up more than 10% of product content. A written statement that the food has been irradiated must be either printed on the package or displayed next to the irradiated food.

Health Canada completed the scientific review of proposed uses of the food irradiation. Related proposed amendments to the Food and Drug Regulations were published in the Canada Gazette, Part I, on November 23, 2002 (Table 12.6).

Europe

European clearances include the following (adapted from IFST 2006):

- Low-dose (less than 1 kGy) irradiation is adequate for insect control (for instance in grain and grain products) where a dose of 150–700 Gy is sufficient.
- Doses of up to 3 kGy (fresh) and up to 7 kGy (frozen) have been recommended for poultry and poultry products, including mechanically recovered meat, to reduce numbers of Salmonella, Campylobacter and other food poisoning bacteria.
- Doses of up to 4.5 kGy (fresh) and up to 7 kGy (frozen) have been recommended for red meats, including particularly hamburger meat, to reduce numbers of *E.coli* O157:H7 and other food poisoning bacteria.
- Doses up to 10 kGy have been recommended for dried herbs and spices to reduce levels of contaminating microorganisms generally and to reduce or eliminate food poisoning bacteria in particular. Alternative methods to reduce microbial numbers have used chemicals, such as ethylene oxide and methyl bromide that are now considered dangerous to humans and/or the environment. Steam flash pasteurization has been another technique for spices.
- Doses up to 3 kGy have been recommended for some seafood, in particular warm water shrimp/prawns and other shellfish, to improve their microbiological safety. Lower doses (< 3 kGy) eliminate 90-

TABLE 12.6. Permitted Use of Irradiation in Canada.

Item	Column I Food	Column II Permitted Sources of Ionizing Radiation	Column III Purpose of Treatment	Column IV Permitted Absorbed Dose
1.	Potatoes	Cobalt-60, Cesium-137, X-rays generated from a machine source (5 MeV max.) or electrons generated from a machine source (10 MeV max.)	To inhibit sprouting during storage	0.15 kGy max.
2.	Onions	Cobalt-60, Cesium-137, X-rays generated from a machine source (5 MeV max.) or electrons generated from a machine source (10 MeV max.)	To inhibit sprouting during storage	0.15 kGy max.
3.	Wheat, flour, whole wheat flour	Cobalt-60, Cesium-137, X-rays generated from a machine source (5 MeV max.) or electrons generated from a machine source (10 MeV max.)	To control insect infestation in stored food	0.75 kGy max.
4.	Dehydrated seasoning preparations	Cobalt-60, Cesium-137, X-rays generated from a machine source (5 MeV max.) or electrons generated from a machine source (10 MeV max.)	To reduce microbial load	10.00 kGy max. Total overall average dose
5.	Mangoes	Cobalt-60, Cesium-137, X-rays generated from a machine source (5 MeV max.) or electrons generated from a machine source (10 MeV max.)	To control insect infestation during storage and to extend durable life	0.25 kGy min., 1.5 kGy max
6.	Fresh poultry	Cobalt-60, Cesium-137, X-rays generated from a machine source (5 MeV max.) or electrons generated from a machine source (10 MeV max.)	To control pathogens, reduce microbial load and extend durable life	1.5 kGy min., 3.0 kGy max.
7.	Frozen poultry	Cobalt-60, Cesium-137, X-rays generated from a machine source (5 MeV max.) or electrons generated from a machine source (10 MeV max.)	To control pathogens, reduce microbial load and extend durable life	2.0 kGy min., 5.0 kGy max.
8.	Fresh, prepared or dried shrimp and prawns	Cobalt-60, Cesium-137, X-rays generated from a machine source (5 MeV max.) or electrons generated from a machine source (10 MeV max.)	To control pathogens, reduce microbial load and extend durable life	1.5 kGy min., 3.0 kGy max.
9.	Frozen shrimp and prawns	Cobalt-60, Cesium-137, X-rays generated from a machine source (5 MeV max.) or electrons generated from a machine source (10 MeV max.)	To control pathogens, reduce microbial load and extend durable life	1.5 kGy min., 5.0 kGy max.
10.	Fresh ground beef	Cobalt-60, Cesium-137, X-rays generated from a machine source (5 MeV max.) or electrons generated from a machine source (10 MeV max.)	To control pathogens, reduce microbial load and extend durable life	1.5 kGy min., 4.5 kGy max.
11.	Frozen ground beef	Cobalt-60, Cesium-137, X-rays generated from a machine source (5 MeV max.) or electrons generated from a machine source (10 MeV max.)	To control pathogens, reduce microbial load and extend durable life	2.0 kGy min., 7.0 kGy max.

95% of spoilage organisms, resulting in an improvement in shelf-life and an elimination of all vegetative bacterial pathogens. Shrimp in ice have a shelf life of 7 days; treating with 1.5 kGy adds another 10 days. 1 kGy eliminates both *E. coli* and *Vibrio spp.* in oysters without detracting from their raw quality. Oyster meats treated with 2 kGy have a shelf-life of 21 to 28 days under refrigeration, compared to 15 days for their unirradiated counterpart.

- Doses of up to 2 kGy have been recommended for certain fruits and vegetables in order to reduce the number of microorganisms, particularly those that cause spoilage. Irradiation has been shown to have minimal effect on flavour, aroma and colour but can have an adverse effect on texture. Onions, garlic, mungbeans and tamarind are commercially irradiated. Irradiation also is useful in combating rice weevil (*Sirohilus oryzae*) and lesser grain borer (*Rhyzopertha dominice*).
- Doses of less than 1 kGy have been recommended for bulbs and tubers, such as potatoes and onions, to prevent sprouting.
- Doses of 1 kGy are recommended for cereals, grains and certain fruits, such as papaya and mango as a quarantine measure, to kill insects.
- High-dose irradiation produces sterile foods, such as ready to eat meals, for special medical diets, emergency or space diets. These foods are irradiated by doses of 20–45 kGy to render the foods commercially sterile. The irradiation is carried out under frozen conditions to minimize adverse sensory effects. The foods can be subsequently distributed unrefrigerated.

ECONOMIC ASPECTS

The beneficial effects of irradiation of fruits, such as disinfestation, reduction of spoilage and extension of shelf life, offer potential for significant cost savings. Due to the limited application of irradiation, the economical feasibility of food irradiation has to be examined on a case-by-case basis. Average costs per pound of irradiating food are similar for the electron accelerator and cobalt-60 irradiators analyzed in this study, but initial investment costs can vary by $1 million. Irradiation costs range from 0.5 to 7 cents per pound and decrease as annual volumes treated increase. Cobalt-60 is less expensive than electron beams for annual volumes below 50 million pounds. For radiation source requirements above the equivalent of 1 million curies of cobalt-60, electron beams are more economical (An Economic Analysis of Electron Accelerators and Cobalt-60 for Irradiating Food, Rosanna Mentzer Morrison, United States Department of Agriculture, Economic Research Service Technical Bulletin Number 1762). According to another report, in a typical food irradiator (GRAY*STARTM FOOD IRRADIATOR) operated at full capacity, 24 hours a day, 350 days per year, the volume processed would be 95.7M lb at less than one half cent per lb. The cost of irradiating various foods was estimated between $0.02 US and $0.40 US per kg (WHO, 1988). According to another design (i.e., a 60 Mkg per year, 0.15 kGy and 0.4 g cm^3), unit processing costs would vary from below $0.01 per kg for 250 Mkg per year to $0.107 per kg for 10 Mkg per year. The fact that more commercial irradiators are appearing in the market indicates a better potential for the economic viability of the technology.

THE FUTURE

Irradiation applied to fruits, vegetables, nuts and spices is increasingly recognized as an effective method for reducing postharvest losses, ensuring hygienic quality of produce and facilitating international trade of particular exotic fruits. Progress in the commercial use of food irradiatiaon has been slow, but there have been positive signs along the way. For this technology to be successful, the application should be filling a real need. The irradiation processes should either be the only solution to a specific problem or possess real advantages over existing technologies. The cost should be comparable to other processes. The trend in the practical application of irradiation for fruits is likely to increase in the coming years in view of restrictions on chemical fumigants used for insect disinfestation. Because irradiation can be applied to fruits in a "fresh" like state and it can kill microbial contaminants and can sterilize or kill adult insects as well as larvae and eggs, it is an alternative process of considerable interest. In developing countries, postharvest food losses are enormous. A low-dose application of irradiation (up to 1 kGy) to fruits at postharvest offers a unique opportunity to eliminate insect infestation. Developing countries will likely benefit more from this technology by reducing their losses and increasing the supply of certain produce. Consumers will benefit from greater price stability due to the availability of many commodities, including tropical fruits throughout the year.

CHAPTER 13

Postharvest Pathology

INTRODUCTION

AMONG the factors in postharvest spoilage of fruits and vegetables previously discussed are many biochemical and physiological causes. Also presented were various means of controlling them for extension of the postharvest storage. The various factors that influence different biochemical and physiological reactions also were discussed, as well as how many of these factors could be simultaneously managed in controlled and modified atmosphere storage, thereby significantly improving the postharvest quality and shelf life of perishable produce. Another mode of spoilage of tender crops is by the attack and growth of microorganisms, one of the most obvious and common causes of postharvest spoilage, especially in the field. Microorganisms continue to cause damage during storage, but their intensity is much less due to the controlled conditions that exist under storage conditions. Not all microorganisms are of concern. Those responsible for attacking produce and subsequently making them succumb to intense spoilage are called "the plant or postharvest pathogens." Although they are pathogenic to plant tissues and cause extensive damage, not all are harmful to humans.

Economic losses caused by postharvest diseases generally are much greater. This is mostly as a result of various activities added in the later part of the postharvest management chain to increase the market value and potential of the crop. Hence, a 10% loss in the field may constitute only a small fraction of the total loss as compared to microbial spoilage at the end of the cycle, since the cost of packinghouse preparations, packaging, cooling, cold storage, transportation and handling are added to the end product with a profit mark-up at every stage. Postharvest diseases affect almost all crops, and the incidence is especially high in developing or underdeveloped countries, which often lack adequate facilities for the proper handling and storage of the produce. Developed countries are not exempt, since even under controlled conditions, pathogenic activity continues along with all other life-sustaining activities.

Specific causes of postharvest losses of fruits and vegetables may be classed as pathogenic, non-pathogenic or physical. This chapter deals with the parasitic or pathological activity of bacteria and fungi. The infections generally occur in the field and manifest themselves during the postharvest chain. Microbial infections can stay in the latent or dormant state for extended periods of time and lead to full-blown infections at a later stage when conditions become favorable for their growth and activity. Fungi are more commonly found attacking fruit, and bacteria are more common as postharvest pathogens in vegetables.

COMMON DISEASES

There are thousands of fungi and hundreds of viruses and bacteria that can attack plants and plant products. So it is important to understand disease development and to develop control methods to reduce their incidence and activity. Despite diversity in these organisms, most diseases are caused by species of just a few genera. The common fungi belong to genera: Alternaria, Botrytis, Colletotrichum, Diplodia, Monilinia, Penicillium, Rhizopus, and Sclerotinia. Pseudomonas and Erwinia are the most common bacteria associated with postharvest spoilage of most vegetables. Many of the common types of storage diseases are named after the color of the mycelium they spread or the type of wound they inflict: grey mold rot (by Botrytis on pome, stone fruits, berries, kiwi, beans, cabbage, carrot), blue/green mold rot (Penicillium on pome, stone fruits, citrus, onion), brown rot (Monilinia on stone fruits, citrus), stem end rot (Diplodia and Alternaria on citrus), watery rot (sclerotonia on carrot, lettuce, celery), soft rot (Penicillium, Erwinia and Pseudomonas on a majority of fruit and vegetables). Others are named after the causative microorganisms: Rhizopus rot (stone fruits, strawberries, papaya, tomato), Alternaria rot (pome and stone fruits, citrus, papaya, tomato), Diplodia rot (citrus, avocado), Fusarium rot (pineapple, banana), etc. Excellent descriptions of postharvest diseases are given in the ehow website: http://www.ehow.com/list_6772603_ postharvest-diseases-fruits-vegetables.html.

Brown rot: This is a fungal disease that may cause serious damage to stone fruits during wet seasons. *Monilinia fructicola* is one of several species of Monilinia that cause brown rot. Prolonged wet weather during bloom may result in extensive blossom infection. Early infections appear as blossom blight or twig canker. Later infections appear as a rot of ripening fruit on the tree and in storage. Spring infections arise from mummified fruit of the previous season that remains attached to the tree or has fallen to the ground. Brown rot causes blossom blight, twig blight, twig canker, and fruit rot.

Rhizopus rot: This is a soft fungal rot of harvested or over-ripe stone fruits. Fungal growth and fruit decay are greatly retarded in cold storage but advance rapidly at warm temperatures, allowing loss of many fruits within the shipping container. The lesions are cinnamon or chocolate-colored. Rhizopus rot causes the skin to slip readily from the decaying flesh underneath. After harvest, Rhizopus rot can spread from fruit to fruit without injury at the point of contact.

Diplodia rot: Diplodia ear rot, caused by the fungus *Stenocarpella maydis*, is a common fungal disease in corn. *Diplodia ear* rot is easy to recognize with a grayish or grayish-brown mold on and between the kernels, and usually only on part of the ear. Occasionally, disease symptoms occur only at the tip-end or middle part of the ear. Diplodia stem-end rot is caused by the fungus *Diplodia natalensis*. It is a major decay organism of citrus fruit produced in warm and humid climates. Diplodia stem end rot is rarely observed on fruit attached to the tree, even when they are mature.

Erwinia rot: Bacterial wilt, caused by the bacterium *Erwinia tracheiphila*, is a destructive disease of plants in the cucumber family. Although bacterial wilt is most common on muskmelon and cucumber, it also can infect squash, pumpkins and a number of wild cucurbit plants.

Fusarium rot: Fusarium rot is one of the more common preharvest and postharvest diseases of cucurbit fruits. Symptoms of Fusarium fruit rot vary depending on the *Fusarium* species and the host. *Fusarium solani* causes foot, root, stem, and fruit rot of cucurbits. *Fusarium* fruit rot is the leading cause of cantaloupe fruit losses. Fusarium fruit rot of muskmelon is caused by the soil-borne fungus *Fusarium roseum*.

Anthracnose: This is caused by the fungus *Colletotrichum lagenarium*, which can be a destructive disease of muskmelons during warm, wet growing seasons. The disease also attacks watermelon, cucumber and gourds.

Alternaria rot: This is a fungal infection that begins while fruits or vegetables still are on the vine or tree. The spores come from decaying matter around the fruit or vegetable, and are transported directly to the fruit or vegetable by wind. Alternaria rot mostly affects carrots, broccoli, potatoes, peppers, apples, kiwis, pears and tomatoes. Symptoms of Alternaria rot are black or brown round lesions around breaks in the skin of fruit or vegetable. The rot eventually will continue to the interior of the fruit or vegetable, ruining their potential for harvesting completely. Alternaria fruit rot is a serious fungal disease of navel oranges and lemons that causes extensive damage. Symptoms of infection consist of dark spots on the fruit's skin and rotting on the inside of the fruit to the core. However, symptoms generally are not severe until postharvest, which makes control during the growing season difficult. Alternaria leaf spot, caused by the fungus *Alternaria cucumerina*, affects muskmelon and cucumber as well as other cucurbits.

Soft rot: Bacterial soft rots are caused by several types of bacteria, but most commonly by species of gram-negative bacteria, Erwinia (now Pectobacterium), and Pseudomonas. The bacteria mainly attacks the fleshy storage organs of their hosts (tubers, corms, bulbs, and rhizomes), but they also affect succulent buds, stems and petiole tissues. With the aid of special enzymes, the plant is turned into a liquid mush in order for the bacteria to consume the plant cell's nutrients. Unlike the soft rot caused by Rhizopus, bacterial rot produces an unpleasant smell.

Stem end rot: Stem-end rot in fruits and vegetables is caused by different microorganisms. Diplodia stem-end rot is a postharvest fungal disease. The fungus favors humid weather and rain for infection. While the disease infects the citrus fruit's stem during the growing season, symptoms do not appear until postharvest. Symptoms appear as brown, rotting streaks on both ends of the fruit, with rot occurring beneath the fruit's skin to the core.

Sour rot: Sour rot is particularly acute in citrus produce, and is caused by fungal spores, which thrive in the soil around a citrus tree and are carried by wind onto the fruit. Sour rot can enter fruits only through punctures or openings in the skin. Older fruits are most susceptible to sour rot. Symptoms are similar to blue mold and include watery and brown lesions.

Watery rot: Watery soft rot is caused by the fungus *Sclerotinia tri-*

foliorum. Leaves or petioles become flaccid and light brown in colour. Disease spreads through the plant to neighboring plants. Patches of clover can rot, causing a light brown slimy mass of decaying vegetation. Tufts of white fungus develop on affected tissue.

Blue mold: This is also referred to as Penicillium rot and it is a prevalent postharvest disease among fruits, most commonly apples. The symptoms of blue mold are soft, light brown areas that may be covered in white or light blue spores. Blue mold spreads quickly even in refrigerated containers, and a few infected fruits can spoil an entire harvest.

Grey mold: This is caused by *Botrytis cinerea*, a fungus that affects many plant species, although its most notable hosts may be wine grapes. The fungus gives rise to grey rot as a result of consistently wet or humid conditions, and typically results in the loss of the affected bunches. *Botrytis cinerea* affects many other plants. It is economically important on soft fruits such as strawberries, tomatoes and bulb crops. Unlike wine grapes, affected strawberries are not edible and must be discarded.

In recent years, it has been recognized that metabolites produced by fungi in foods can be *extremely toxic to humans and other animals*. The most notable example is the production by *Aspergillus flavus* infection of nuts and dried fruits of the mycotoxin aflatoxin, one of the most potent carcinogens known. Four major aflatoxins are: B1, B2, G1, and G2. Patulin, a genotoxin, occurs in apple products, is produced by Penicillium expansum and others. Patulin is heat stable and toxic to many biological systems, but its role in causing animal and human disease is unclear. Produced by *Penicillium viridicatum* and *Penicillium ochraceous*, ochratoxin A is a carcinogen, and also a blood, cardiovascular, liver, and kidney toxicant. It occurs in wine grapes and dried grapes.

Contamination of horticultural products by fecal coliforms has been increasing dramatically and has been reported to be responsible for several outbreaks linked to Salmonella, *Escherichia coli* and Listeria. These contaminations may also lead to interactions with other plant pathogens resulting in enhancement of their pathogenic activity.

DISEASE DEVELOPMENT

The common terms and terminology associated with postharvest and plant pathology are given in Table 13.1. An excellent summary of the disease development is provided by Kader (1985). Plant product dete-

***TABLE 13.1. Common Terms and Terminology associated
with Plant Pathology.***

Name	Description
Pathogen	An entity, usually a microorganism that can incite the disease.
Parasite	An organism living on or in another living system (host) obtaining its food from the host.
Infection	Establishment of a parasite (pathogen) within a host plant.
Pathogenicity	The capability of a pathogen to cause disease
Symptom	The external or internal reactions or alterations of a plant as a result of a disease *Chlorosis:* Yellowing of normally green tissue due to chlorophyll destruction or failure of chlorophyll formation *Lesions:* Visible disease area, including, leaf spot, necrotic spot, mosaic
Sign	The pathogen or its parts or products seen on a host plant. *Examples:* Gray mold, Blue mold, Green mold, Powdery mildew on pome, stone fruits, citrus, onion
Disease	Malfunctioning of host cells and their tissues that result from their continued invasion by a pathogenic agent or environment factor and leads to development of decay symptoms.
Rots	The softening, discoloration and often disintegration of a succulent plant tissue as a result of fungal or bacterial infection.

rioration typically results from either bacterial or fungal infection. In order for the microorganisms to attack, they must first gain access to the internal tissues of the fruit or vegetable. They generally do so by one of the two modes of entry: (1) penetrate healthy tissue or (2) enter through wounds and bruises.

The healthy host plants are generally resistant to a vast majority of microorganisms but are susceptible only to a few species and in some cases only to specific strains of certain pathogen species. The protective tissues of healthy plant organs provide an excellent barrier against microbial attack and generally resist invasive pathogens. This defense mechanism is believed to be very strong in growing and healthy tissues, which can resist the spread of disease and suppress a pathogen's activity. Maturity also plays an important role in the defense of the plant organs against microbial attack. The defense, which is very strong in young tissues, continues to persist until the organ matures, after which the resistance begins to drop. Tissues are weakened as the mature fruit or vegetable goes through the stages of ripening and senescence. During these latter stages, the tissue becomes vulnerable to the attack of the same pathogens it resisted when young.

MECHANISM OF PATHOGEN ENTRY

In order to be successful on a host plant tissue, a pathogen must first gain access to the tissue and then overcome the defense responses. In other words, it must be able to attach itself to the plant tissue and then generate necessary resources so that it can penetrate and invade the tissue. During the entire process, it also should be able to overcome the defense response of the host tissue.

THE INFECTION PROCESS

When the pathogen lands and becomes attached to the plant structure, it can stay there in a dormant stage as a spore under unfavorable conditions. When favorable conditions are established (appropriate temperature, available moisture, food, etc.), the spore coat swells and produces a germ tube with the simultaneous secretion of enzymes (cutinase, pectinase) capable of hydrolyzing the epidermal layers. The spore tube develops and ruptures the softened skin tissue, eliciting a strong defense reaction from the plant tissue. The germinated spore may die if it cannot overcome this defense response. If it can overcome the defense, further development takes place and the pathogenic fungi will spread its mycelium and invade the plant tissue. More enzymes are secreted, which eventually kill the plant cell. The same series of reactions can take place when the spores land on wounds. If they can overcome the initial defense, further development takes place. Once well established in a primary host, the mycelia of some fungi can invade the healthy tissues of adjoining fruits and vegetables via stomates, lenticels, or sometimes directly via the epidermis, leading to "nesting" where several adjoining commodities in a package are destroyed by the fungus.

DEFENSE MECHANISM

Increased attention has been given in recent years to studying the exact nature of the defense mechanism and the functions of various components involved in it. It is believed that as soon as a pathogen lands on the tissue, there will be a recognition response to find out whether the invasion is natural and part of its own growth and development sequence (self-recognition) or the invasion is external (non-self invasion). In the event of non-self entry, the recognition process will continue to "see" whether the attack is species-specific or nonspecific, which may be important in the type of defense response. The message is then transferred to the nucleus eliciting the appropriate response. The defense response then follows to counteract the invasion. The defense response is termed *hypersensitivity* or *hypersensitive reaction*, which has been found to be associated with the synthesis of phenolic compounds with terpenoid and flavonoid structures. These low molecular compounds possess antimicrobial properties and are commonly called phytoalexins (or stress metabolites). Some of the phenolic compounds are naturally present in the tissue and also are formed in large quantities in response to a pathogen attack. They are formed from polyphenolic substrates by the action of the enzyme, polyphenol oxidase. Both the enzyme and the substrate exist naturally in the plant tissue; however, only when the tissue is stressed (ruptured, broken down or invaded) are the enzyme and substrate brought together. Other enzymes such as pectin methyl esterase (hydrolyzing the methyl esters in pectic acid) and polygalacturonase (hydrolyzing pectic acid to galacturonic acid) also are implicated. The oxidative reactions lead to brown pigments. These reactions in the ruptured cells are believed to activate neighboring cells and initiate the repair mechanism. The reactions have also been implicated in the wound-healing process in which the plant tissue tries to develop new periderms to patch up the bruises caused by external agents.

GROWTH BEHAVIOR OF BACTERIA AND FUNGI

Once settled on the host, microorganisms will start to grow, provided conditions are favourable, i.e., they have all the food and other requirements for their support. Microbial growth under favorable conditions generally is expected to follow a sigmoid curve (Figure 13.1) demonstrating a lag phase, a log phase and a stationary phase, which can be followed by a phase of death and population decline. In the lag phase, the microbe will try to establish and condition itself for growth in the medium. As it leaves the lag phase, it will continue to grow at a faster rate leading to an exponential (logarithmic) or accelerated growth phase. This represents the active stage. Then, while they still continue to grow in large numbers, part of the microbial population will begin to die, thereby moderating the growth. The net result is a decline in the rate of growth. Eventually, the multiplication and decline reach a state of equilibrium, which results in the stationary phase. When they cannot sustain this, the population begins to decline, which may happen due

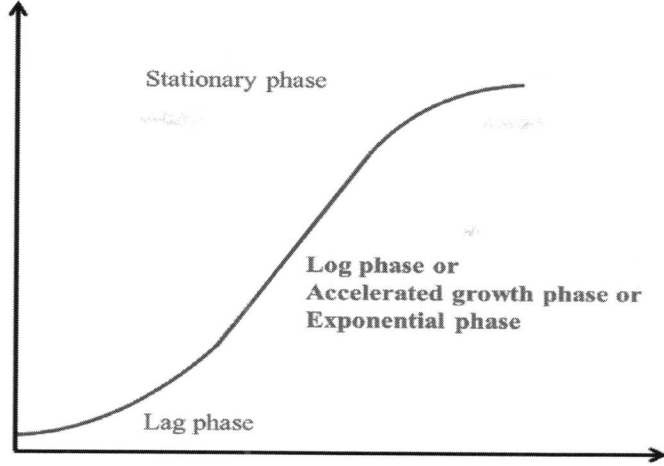

FIGURE 13.1. The microbial sigmoid growth curve.

to exhaustion of vital supplies or due to strong competition developing within the environment.

The microbial growth pattern is influenced by many factors. Favorable conditions shorten the lag phase and accelerate the log phase, while unfavourable conditions help to extend the lag phase and also reduce the rate of multiplication of cells in the log phase as shown in Figure 13.2. Factors that contribute to accelerated disease growth include higher temperature and higher relative humidity, delayed or slower rate

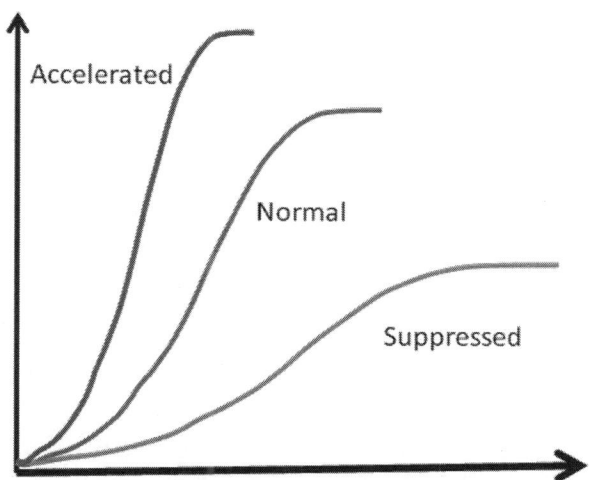

FIGURE 13.2. The microbial sigmoid growth curve.

of cooling and improper storage conditions, etc. On the other hand, all control measures that are taken to reduce the postharvest physiological activities like temperature control, humidity control, controlled and modified atmosphere storage and packaging conditions, presence of antimicrobial agents, etc., help to delay the growth and activity of the pathogens and therefore help to prevent spoilage.

TREATMENTS FOR PATHOGEN CONTROL

Fungicides

Although small as compared to pre-harvest pest control products, many fungicides are used for control of postharvest diseases caused by microorganisms. Residue concerns, toxicological effects and unfavorable influences on nutritional and sensory quality have resulted in abandoning many chemical fungicides and bactericides. Others have lost importance due to development of pathogen resistance to the fungicides. Examples of postharvest chemical treatments that are presently used are thiabendazole, dichloran and imazalil. Use of preservatives or antimicrobial agents generally are not recognized as postharvest treatments but sometimes they help control postharvest decay of fruits and vegetables. Common preservatives for food applications include sodium benzoate, SO_2, potassium meta bi-sulfite, potassium sorbate, sodium propionate, acetic acid, nitrites and nitrates, chitins, chitosans and some antibiotics like nisin. SO_2 is exclusively used for grapes.

Biological Control

Biological control of weeds in pre-harvest applications has become extremely popular; however, its use in postharvest applications is still in its infancy. Several biological control agents have been developed in recent years. A strain of *Pseudomonas syringae* was found to control Blue and Grey mold of pome fruit (Janisiewicz and Marchi, 1992). Other bacterial microorganisms are being developed for postharvest disease control. For example, strains of *Bacillus pumilus* and *Pseudomonas fluorescens* have been identified that exhibit successful control of *B. cinerea* in field trials of strawberries (Swalding and Jeffries, 1998). Yeasts such as *Pichia guilliermondii* (Wisniewski et al., 1991) and *Cryptoccocus laurentii*, a yeast that occurs naturally on apple leaves, buds, and fruit (Roberts, 1990) were the first to be applied for control of postharvest decay on fruit. The yeast *Candida oleophilia* has been registered for control of postharvest decay on fruit crops. Although biological control methods are promising, they may not show consistent results because their efficacy is affected by the amount of pathogen present, compatibility with other chemicals used, disease adaptation, etc.

UV and Gamma Radiation

Although ultraviolet light has been well recognized as an effective treatment of bacteria and fungi on food contact surfaces, it has not been shown to be effective for reducing decay of pre-packaged fruits and vegetables. Low doses of ultraviolet light irradiation (254 nm UV-C) have been reported to reduce postharvest brown rot of peaches (Stevens *et al.*, 1998). In this case, the low-dose ultraviolet light treatments were shown to be effective on brown rot development because of two effects: reduction in the inoculum of the pathogen and induced resistance in the host. However, it has not become a practical postharvest treatment as yet and requires more research. Recent research has shown evidence of low-dose UV irradiation to induce stress metabolism so as to improve the antioxidant properties by enriching polyphenolics.

Irradiation has been discussed in a separate section (Chapter 12). Gamma radiation has been well recognized for controlling decay, disinfestation, and extending the storage and shelf-life of fresh fruits and vegetables. Low doses of 1–2 kilogray (kGy) have been effective in controlling decay in several fruits and vegetables. Commercial application of gamma radiation is expanding in spite of increased consumer resistance. One of the most attractive incentives for the low dose gamma radiation for fruits and vegetables is the possibility of eliminating chemical fumigants for infestation control. Current chemical fumigants such as ethylene di-bromide and methyl bromide are being discontinued to prevent damage and to protect the ozone layer surrounding the earth.

Effect of Storage Environment

Since the growth and activity of bacteria are dependent on environmental temperature, relative humidity and air composition, these factors should be maintained at levels that are most conducive to suppressing bacterial and fungal growth/activity. Temperature management is critically important in postharvest disease control, and all other treatments can be considered as supplements to refrigeration. As discussed earlier, fruit fungi generally grow optimally at 20 to 25°C, and their growth can be suppressed significantly at temperatures between 0 and 5°C. There are some fungi which can grow at temperatures below –2°C and not much can be done to retard them, unless frozen storage can be adopted. As with respiratory and physiological activities, temperatures as low as possible should be used to reduce the growth and activity of bacteria and fungi.

In terms of relative humidity, both low and high levels have implications with respect to postharvest decay control. Perforated polyethylene bags are commonly used for fruit and vegetable storage in order to create high relative humidity so that transpiration resulting in weight loss, shriveling, etc., is reduced. Such high relative humidity environments are conducive to mold growth; it is important that surface moisture on produce be eliminated. Moisture can appear on produce surfaces due to condensation when the water vapor pressure deficit is negative (produce surface temperature below the dew point temperature of the air) or due to condensation when the water vapor pressure deficit is positive but the air is saturated. Both these situations were discussed earlier in the section of transpiration. Crops with a thick skin, peel, and well-developed cuticle and epidermis can tolerate lower RH levels that help prevent storage decay. Fungal spore germination also can be considerably inhibited at low relative humidity, which can aid in extending the shelf-life. Protective coatings with or without fungicides can also reduce the incidence of postharvest decay.

Atmospheric alterations of reduced O_2 and elevated CO_2 levels are often provided around fruit and vegetables either within packages (modified atmosphere packaging) or in storage rooms (CA storage) in order to reduce produce respiration. Since pathogens also need to respire in order to survive and show activity, lowering the O_2 concentration to around 2–3% or raising the CO_2 above 5% can help to suppress pathogenic activity in the produce. CO_2 added to air has been widely adapted in the transportation of strawberries and cherries, primarily to suppress mold growth. Higher levels of CO_2 above 5% are more effective than lower levels of O_2 since even 2% oxygen concentration can sustain mold growth. Hypobaric storage in which O_2 levels close to 1% are maintained can be effective against mold growth.

Hot water and vapor heat treatments may be used for effective control of pathogens on crops such as mango, papaya, pepper and tomato. Likewise, cold treatments can also be effective. These are described in the next chapter.

PRE-HARVEST CONTROL

Pre-harvest treatments can help to control postharvest problems. Sometimes, latent infections can develop on the senescing flower parts and in such cases fungicide sprays during the flowering period may provide excellent control of postharvest diseases, as demonstrated by control of Botrytis rot on strawberries and raspberries. The following are examples of orchard control pre-harvest sprays which may have ongoing fungicidal or fungistatic effects:

- Aromatic amines—Rhizopus, Monilinia and Botrytis—Stone fruits, carrot
- Chlorine—bacteria & fungi in wash water—produce
- Formaldehyde—fungi—packages
- SO_2 fumigation—Botrytis—grapes
- SOPP (Na o-phenyl phenate)—Penicillium—citrus, apple
- Benomyl—Penicillium, Monilinia, Botrytis—apple, citrus, peaches, apricot
- Thiabendazole (TBZ)—Diplodia, Penicillium, Botrytis—avocado, citrus, kiwi
- Imazalil—Penicillium—citrus, pineapple
- Dichloran—Rhizopus, Monilinia, Botrytis—stone fruits, carrot, sweet potato

CHAPTER 14

Postharvest Treatments

INSECT CONTROL

INSECT infestation is one of the major causes of postharvest losses of staple foods in tropical countries. With fresh fruits and vegetables, the bulk of the damage by insects and pests comes when the produce is in the field or orchard. Several species of insects with common names, such as fruit flies, weevils, moths, mites, bugs, etc., breed on horticultural produce.

Although the actual incidence of postharvest spoilage of fruits and vegetables by insects is not alarmingly large, fruits and vegetables carry eggs and larvae to wherever they are shipped. The insect eggs can stay in a dormant state for long periods, sometimes a whole season if the commodity is stored that long. The eggs can transform themselves to insects any time when favorable conditions exist.

Therefore produce acts as carriers of potentially harmful insects and pests, which can infiltrate the various geographical areas along their shipping route, causing damage to farms and orchards. Generally, certain insect species are dominant in certain geographical areas and breed on typical horticultural crops. For example, the Mexican fruit fly (Figure 14.1) breeds on citrus and other tropical/subtropical fruits, distributed mainly in Central America and Mexico. Likewise, the apple maggot, which breeds on apple crop, is especially dominant in the USA and Canada. The Mediterranean fruit fly (Figure 14.1), sometimes referred to as medfly, is a widely distributed fruit fly that breeds on deciduous and subtropical fruits. It is one of the most widely distributed insects

FIGURE 14.1. Mexican (a) and Mediterranean (b) fruit flies.

in the world—breeding across South and Central America as well as Africa.

In order to prevent the spread of insects to other geographical areas, many countries place stringent import control procedures. This requires treatments to be given to produce originating from locations suspected of widespread distribution of harmful insect species. These mandatory treatments set up by the importing countries for insect control are called "Quarantine Treatments for Insect Control". It is up to the exporting countries to meet all the quarantine requirements of the importing country for establishing continued produce trade links. Many times, these quarantine requirements or trade barriers as some call them, can cause serious disruption of produce trade between two countries or in some cases even between two states or provinces within the countries.

In most countries, arriving passengers have to complete a declaration form in which they declare they are not bringing any agricultural commodity, that they have not recently visited any agricultural farm and are not planning to visit any farm immediately after arrival. A typical Canadian Customs Declaration form is shown in Figure 14.2. These are some steps the government takes to ensure the safety of the farmland. Such restrictions also are placed on movements from one province to another province or one state to another state within a country. These procedures safeguard farmlands in terms of reducing the risk of insect contamination.

Agriculture and Agri-Food Canada and Canadian Food Inspection Agency (CFIA) declare: *"You can prevent foreign pests and diseases from damaging or destroying Canada's animals' fields, farms and forests. Agricultural products such as meats, plants, seeds and soil must be inspected by Agriculture Canada to ensure that these products are free from harmful pests and viruses. Diseases brought into Canada through agricultural items could cost Canada billions of dollars and increase the cost of agricultural goods to the Canadian consumer. Please help us prevent the entry of these pests and diseases into Canada by declaring all animal and plant products on the declaration card. The following items commonly carried by travelers must be declared: meat and meat products; cream, milk, cheese and other dairy products; plants, trees, cut flowers and their soil (may require an import permit); wood and wood products; fruits and vegetables (may require an import permit); pets, birds and other live animals (require an import permit or vacci-*

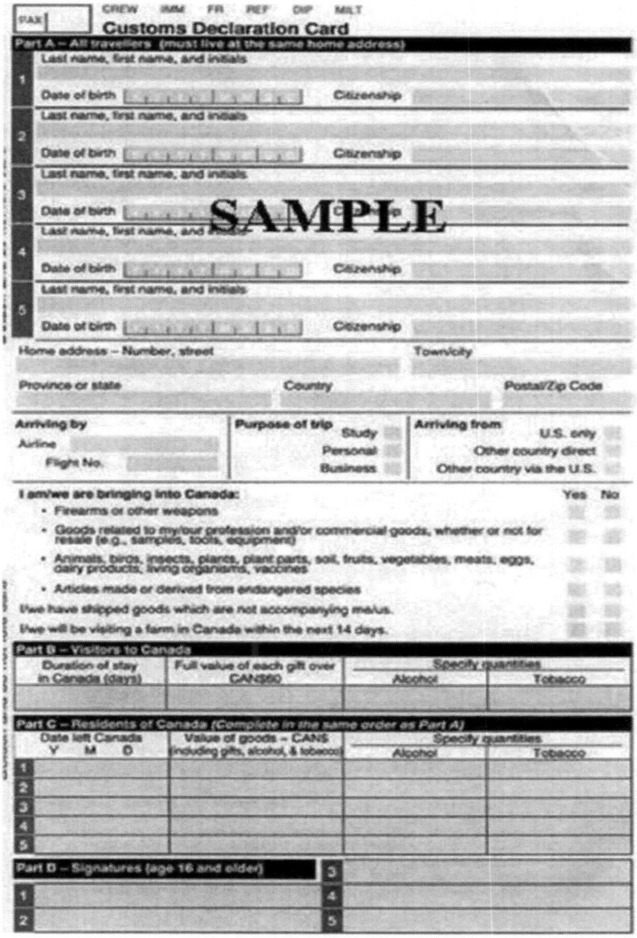

CREW IMM FR REF DIP MILT

Customs Declaration Card

Part A – All travellers (must live at the same home address)

1 Last name, first name, and initials
Date of birth Citizenship

2 Last name, first name, and initials
Date of birth Citizenship

3 Last name, first name, and initials
Date of birth

SAMPLE

4 Last name, first name, and initials
Date of birth Citizenship

5 Last name, first name, and initials
Date of birth Citizenship

Home address – Number, street Town/city
Province or state Country Postal/Zip Code

Arriving by: Airline, Flight No.
Purpose of trip: Study, Personal, Business
Arriving from: U.S. only, Other country direct, Other country via the U.S.

I am/we are bringing into Canada: Yes No
- Firearms or other weapons
- Goods related to my/our profession and/or commercial goods, whether or not for resale (e.g., samples, tools, equipment)
- Animals, birds, insects, plants, plant parts, soil, fruits, vegetables, meats, eggs, dairy products, living organisms, vaccines
- Articles made or derived from endangered species

I/we have shipped goods which are not accompanying me/us.
I/we will be visiting a farm in Canada within the next 14 days.

Part B – Visitors to Canada

Duration of stay in Canada (days)	Full value of each gift over CAN$60	Specify quantities: Alcohol	Tobacco

Part C – Residents of Canada (Complete in the same order as Part A)

	Date left Canada Y M D	Value of goods – CAN$ (including gifts, alcohol, & tobacco)	Specify quantities: Alcohol	Tobacco
1				
2				
3				
4				
5				

Part D – Signatures (age 16 and older)
1, 2, 3, 4, 5

FIGURE 14.2. *Canadian customs declaration form.*

nation documentation); feathers and down; seeds and nuts; and baby formula." This does not mean you cannot bring such products to a foreign country. AAFC-CFIA also states that *"To import these goods you will need certificates or permits. Many goods do not need mandatory inspection by the CFIA. There are some plants, vegetables and organic material that can be brought into Canada without permits in limited quantities. If you plan to bring anything into Canada when you visit, please check with the Canadian Consulate first."* Such practices exist in almost all countries to protect farmland.

The quarantine treatment is a physical or chemical treatment given to a commodity to ensure the complete destruction of insects at all stages of their growth: eggs, larvae, pupae and adults.

Four techniques have been recognized by the Quarantine Authorities in US and Canada:

- Fumigation
- Cold treatment
- Heat Treatment
- Irradiation

Fumigation

Fumigation is one of the early and most popular methods for insect control. It is a process of subjecting produce to gaseous fumigants in gas-tight enclosures under controlled conditions of temperature, time and concentration. These can be given in specially constructed fumigation chambers with appropriate temperature control. Sometimes, they also are given in emergency makeshift tarpaulin tents or gas tight rail cars or road trucks. No matter which method is used, the procedures have to be given by specially trained operators under the appropriate supervision of the quarantine authorities. In principle, it is no different from the ethylene ripening procedure, which we looked at earlier, except that you are looking at a slightly modified procedure with a different gaseous chemical.

The concentration of the fumigant and duration of fumigation will depend on:

1. Insect species.
2. Degree of contamination.
3. Produce compatibility.
4. Quantity of produce.

Some common fumigants are (EPA, USA):

Ethylene dibromide (EDB): Ethylene dibromide is a colorless liquid with a mild sweet odor, like chloroform. It also is known as 1,2-dibromomethane. The chemical formula for ethylene dibromide is $C_2H_4Br_2$. Ethylene dibromide is slightly soluble in water. Ethylene dibromide has been used as a fumigant to protect against insects, pests and nematodes in citrus, vegetable and grain crops. In 1984, EPA (USA) banned its use as a soil and grain fumigant.

Methyl bromide (MB): Methyl bromide (MeBr) is an odorless, colorless gas that has been used as a soil fumigant and structural fumigant to control pests across a wide range of agricultural sectors. Because MeBr depletes the stratospheric ozone layer, the amount of MeBr produced and imported in the U.S. was reduced incrementally until it was phased out in January 1, 2005 (pursuant to *Montreal Protocol on Substances that Deplete the Ozone Layer*) (EPA, USA).

Phosphine: Phosphine or hydrogen phosphide (PH_3) is a low molecular weight, low boiling point compound that diffuses rapidly and penetrates deeply into materials, such as large bulks of grain or tightly packed materials. The gas is produced from formulations of metallic phosphides (usually aluminium or magnesium phosphide) that contain additional materials for regulating release of the gas. Phosphine is very toxic to all forms of animal life, hence exposure of human beings even to small amounts should be avoided. Phosphine ranks as one of the most toxic fumigants of stored product insects. Because phosphine is highly toxic, inhalation of even small quantities of the dust from the formulation and residual after treatment should be avoided. On completion of the fumigation, all windows and doors should be opened and the space aerated for at least two hours (FAO).

Ethylene Oxide: As an insecticide, the principal use of ethylene oxide (ETO) has been for fumigation of bulk grain in recirculating systems and in the vacuum fumigation of packaged foods and tobacco. Ethylene oxide is flammable within wide limits. It is therefore necessary in many commercial applications to mix it with a nonflammable carrier. It is obtainable mixed with carbon dioxide in the proportion of one part ETO to nine parts CO_2. Despite a general impression to the contrary, ETO must be regarded as poisonous to humans by inhalation, although it is not as lethal in comparatively low concentrations as some other fumigants. Of the commonly used fumigants, ethylene oxide is about intermediate in toxicity to insects (FAO).

Many fumigants have been used in the past. Several are banned today in some countries but are still being used in others.

Cold Treatment

Cold treatment exposes the commodity to a given low temperature long enough to cause destruction of insect life. The treatment is effective against insects of tropical origin, which do not tolerate low temperatures for longer periods. However, the produce itself must be tolerant of the temperature regime for the required duration. These procedures are not appropriate for chilling sensitive commodities.

The following combinations are recognized for treatment of medfly (Kader, 1985):

- 10 days at 32°F or below
- 11 days at 33°F or below
- 12 days at 34°F or below
- 14 days at 35°F or below
- 16 days at 36°F or below

Higher temperatures result in lower effectiveness. Strict control of temperature and time is required. Produce has to be held in quarantine prior to shipment. The duration of 10–14 days may be too long for some commodities. Unless the storage life of the produce is more than a month and it is not chilling sensitive, this procedure is of limited use.

A combination of fumigation and cold treatment may be an alternative. Low-dose fumigation can be combined with the cold treatment to reduce the treatment by a few days.

Heat Treatment

Heat treatment for insect control was practiced for many years. However, in the past 20 years, development of effective fumigants for both insect control and mold growth produced much cheaper, effective and faster methods for insect control. So interest in heat treatment waned. But in recent years as customers became more and more wary about the use of chemicals in foods, more substances are being banned because of their link with carcinogenicity. Hence, alternate disinfestation methods are once again being looked upon with interest. It is from this perspective that heat treatment is enjoying renewed recognition.

Vapor Heat

It has been recognized for more than half a century that Mediterranean fruit flies can be killed by gradual holding of the produce at 40–45°C for 8 hr in humidified warm air. This procedure was called vapor heat treatment. The effectiveness of vapor heat, which is heating of the produce in temperature-controlled chambers with humidified air, also has been recognized as effective against other insects such as the oriental fruit fly, melon fly and a variety of other insects and mites. The exact temperature and time of exposure are dependent on the type of insect and the commodity. Different procedures follow.

In one example, citrus fruits are subjected to 43°C vapor heat for an 8 hr approach time and an additional 8 hr treatment time. In another example, the commodity is quickly heated to 43°C and then held at 47°C for a slower approach, and the treatment was terminated when the pulp reached 47°C. The complete treatment required only 6 hr. For papayas, a 9–13 hr approach time was used at 44°C and a treatment time thereafter for 8.75 hr. Most of these procedures are established by empirical means. Overall, the temperatures used are in the 44–48°C range, and the time varies from 6–12 hr.

Hot Water Dip

Hot water is more effective in terms of heating the produce than vapor heat with humidified air, because water is denser than air and the heat transfer coefficients associated with water are several orders of magnitude higher than those associated with air. Direct steam has the best heat transfer capabilities but its temperature is too high for this type of treatment. Once steam is mixed with air as is used in vapor heat treatment, its heat transfer capabilities are limited because air is a medium with a low heat transfer coefficient, which also interferes with the condensation of steam on the produce surface. It is also easier to control water temperature using thermostatic heaters and the temperature distribution can be much more uniform.

Hot-water treatment has been widely used for several fruit crops such as banana, papaya, mango, melons, etc. Here again, time and temperature are dependent on the commodity and the insect type. For example, bananas have been dip treated in water at 50°C for 20 min. For some fruits, a double dip technique is employed to guard against damage from the heat treatment itself. In this procedure, the produce is first slowly warmed at a slightly lower temperature and then subjected to the actual heat treatment. For example, papyas are frequently subjected to a double dip technique for insect control. For this, they are preheated at 42°C for 30 min followed by a 20 min treatment at 49°C.

Hot water dips have been shown to be beneficial even from a postharvest disease control point of view. Hot water dip treatments have been shown to be effective against: Erwinia, Diplodia, Botrytis, Penicillium, Rizopus and Monilinia. In several instances, fungicides also are added to the water during the treatment. The effectiveness of the fungicide treatment is enhanced at the elevated temperatures. In addition, fungicides can be added to the waxes used on selected fruits and vegetables.

The primary disadvantage of a widespread use of heat treatment techniques is the sensitivity of many fruits to temperatures required for effectiveness. For example, navel oranges, lemons, avocados, peaches, pears, etc., are quite heat sensitive.

CHILLING INJURY

Chilling injury is the permanent and irreversible physiological damage to a sensitive commodity resulting from the exposure of plant tissues to temperatures below a certain critical level. The critical temperature is commodity specific. The general range of temperature in which the produce can undergo chilling injury is 5 to 15°C.

Earlier sections of this book (Chapter 2) display a classification of fruits and vegetables based on their chilling sensitivity. Several factors influence the extent of chilling injury in fruits and vegetable. There also are time limits for the development of chilling injury at any given temperature. For example, at 0°C asparagus can be held for 10 days and watermelon for 7 days before chilling injury can prevail.

It is possible to store any commodity at any temperature for a certain time period without chilling injury. The tolerance of the produce to a given temperature depends on: temperature, duration, commodity, maturity, genetic factors, mode of exposure (continuous or intermittent).

The following are symptoms of chilling injury (Kader, 1985):

1. Surface discoloration, lesions, pitting.
2. Water-soaked areas (common in leaves).
3. Internal discoloration (browning), where surface may look fine.
4. Tissue breakdown and softening.

5. Failure to ripen in the normal pattern.
6. Increased respiration and ethylene producing rates.
7. Increased susceptibility to microbial attack.
8. Accelerated senescence.

A chilling injury can be aggravated by poor handling, low relative humidity storage conditions, and improper air circulation.

Alleviating Chilling Injury

The following procedures have been recognized to be effective for alleviating chilling injury:

1. The best practice is to store the commodities at temperatures slightly above the critical level.
2. *Manipulate the storage environment:* CA conditions can reduce the incidence of chilling injury in some crops such as banana, okra, cucumbers, etc.
3. *Intermittent warming while cooling:* Alternating cooling and warming has been shown to give greater chilling resistance to a commodity. The beneficial effect has been related to restoration of normal metabolic activity during the warm-up period, which enables recovery from the temporary damage caused by the cooling. It also is believed that during this warm-up period, physiologically active components such as ethylene may be removed from the tissue. In order to be effective, the warming-cooling cycle has to be appropriately adjusted to initiate warming before irreversible damage is induced. The temperature in storage can be set cyclically to take advantage of the procedure, as long as the storage is specifically intended for chilling-sensitive produce.
4. The hypobaric storage has also been shown to be effective in terms of lowering temperatures to a certain degree below the critical level.
5. *Chemical treatments (Calcium dip):* Plant tissues with high levels of Ca have been shown to be resistant to chilling injury. Increasing the calcium levels in the plant products either by pre-harvest nutrition or by calcium dip has beneficial effects (e.g., apples, tomato, okra).
6. *Genetic Studies:* Periodically, there have been studies attempt-

ing to breed varieties of certain commodities that have a special resistance to chilling injury. Some studies have shown that seed germination at lower temperatures enables the plant product to have greater resistance to chilling injury. There also have been several tissue culture studies for propagating chilling-resistant varieties. In these, the plant cells from chilling-resistant varieties are isolated and cultured. Some success has been observed with green pepper.

FREEZING INJURY

Commodities stored at temperatures below their freezing points will be damaged. This can occur in the field due to frost, or in storage, handling and transportation due to improper temperature management.

In order for freezing to occur, there has to be nucleation followed by growth of ice crystals, which occurs when the temperature surrounding the produce falls below 0°C. In food products, dissolved solids, impurities and even bacteria can act as active sites for nucleation around which ice crystals can grow. Species of microorganisms such as *Pseudomonas syringae* and *Erwinia herbicola* have been shown to catalyze ice formation at −1°C. Sometimes, it is possible to super cool water below 0°C without freezing it. In an undisturbed setup under carefully controlled temperature, water can be cooled to as low as −10°C. When such a system is given a shock, freezing occurs instantaneously and water converts to ice. If above freezing temperature is restored, there may not be any real freezing.

Commodities also differ with respect to their sensitivity to freezing injury. Asparagus, avocado, beans, cucumber, eggplant, lettuce, okra, pepper, potato, summer squash, sweet potato and tomato are highly susceptible, while broccoli, carrots, cauliflower, celery, onion, parsley, peas, radish, spinach and winter squash are moderately susceptible and beets, Brussels sprouts, cabbage, kohlrabi, parsnip, rutabaga and turnip are least susceptible.

In terms of alleviating freezing injury, there are not many things one can do. Avoiding freezing temperatures and careful control of temperature of the environment are probably the best answers. It is important that large temperature fluctuations in the system be avoided.

Hypobaric and CA conditions may allow the use of temperatures slightly below zero without freezing injury. Use of hypobaric and CA conditions also can permit temperatures slightly above the freezing point without significantly reducing the shelf life. Such conditions will give a greater margin of safety in terms of preventing freezing damage in storage.

References and Reading Materials

Bourne, M. 1977. *Post harvest food losses–the neglected dimension in increasing the world food supply.* Cornell International Agriculture Mimeograph 53. Published by New York State College of Agriculture and Life Sciences, Cornell University.

Brody, Aaron L. 2011. *Modified Atmosphere Packaging for Fresh-Cut Fruits and Vegetables.* New York: Wiley-Blackwell.

Chakraverty, Amalendu and R. Paul Singh. 2014. *Postharvest Technology and Food Process Engineering.* Boca Raton, FL: CRC Press.

Chakraverty, Amalendu, Arun S. Mujumdar and Hosahalli S. Ramaswamy. 2003. *Handbook of Postharvest Technology: Cereals, Fruits, Vegetables, Tea, and Spices.* Boca Raton, FL: CRC Press.

Charm, S. E. 1978. *The fundamentals of food engineering.* Westport, Conn.: AVI Pub. Co.

Codex Alimentarius. 1998. *General standard for contaminants and toxins in foods.* Annex IVB, Waningen, The Netherlands.

Dubuc, B. and C. Lahaie. 1987. *Nutritive Value of Foods.* Montreal: Société Brault-Lahaie.

Duckworth, R. B. 1966. *Fruit and vegetables.* Oxford: Pergamon Press.

European Commission. 2004. 69: European Commission (EC) Scientific Committee on Food (SCF) documents.

FAO. 1989. Prevention of post-harvest food losses fruits, vegetables and root crops a training manual. Rome: Food and Agricultural Organization of the United Nations (FAO UN).

Fellows, P. S. 1988. *Food Processing Technology: Principles and Practice.* Boca Raton, FL: CRC Press.

Fennema, O. R. 1985. *Food Chemistry.* CRC Press, Boca Raton.

Food Irradiation: International Atomic Energy Agency (IAEA) Books. http://www-pub.iaea.org/books/IAEABooks/Subject_Areas/0201/Food-irradiation.

Food Standards Agency (UK). 2002. Food Survey for Irradiated Foods—Herbs and Spices, Dietary Supplements and Prawns and Shrimps. *Food Survey Information Sheet Number 25/02.* June 2002.

Food Standards Agency (UK). 2004. Food Standards Agency Consumer Committee. *Food Irradiation Cons. Comm D030/04.*

IAEA (International Atomic Energy Agency). 1989. Acceptance, Control of and Trade in Irradiated Food. Conference Proceedings. Geneva, December 12–16, 1988, Jointly Organized by FAO, WHO, ITC-UNCTAD GATT, Vienna, Austria.

Institute of Food Science and Technology (IFST). 2006. The use of irradiation for food quality and safety. London: IFST.

Institute of Food Technologists (IFT). 2004. *Irradiation and Food Safety.* Vol. 58, No 11. www.ift.org

International Consultative Group on Food Irradiation. (1991) *Facts about Food Irradiation.* (set of 14 fact sheets covering all aspects of food irradiation issued as public information). ICGFI Fact Series 1-14. IAEA,Vienna.

Janisiewicz, W. J. and A. March. 1992. Control of storage rots on various pear cultivars with a saprophytic strain of *Pseudomonas syringae. Plant Disease* 76: 555–560.

Joint FAO/IAEA/WHO Study Group on High Dose Irradiation (1998). Weekly Epidemiological Record, Jan. 16, 1998.

Kader, A. A. 1985. Postharvest technology of horticultural crops. Regents of the University of California, Division of Agriculture and Natural Resources, Oakland, California 94608. Later version—3rd edition, 2002.

Kader, A. A. 1986. Biochemical and physiological basis for effects of controlled and modified atmospheres on fruits and vegetables. *Food Technology* 40(5): 99–100 and 102–104.

Kester, J. J. and O. Fennema. 1989. An edible film of lipids and cellulose ethers: Barrier properties of moisture vapor transmission and structural evaluation. *J. Food Sci.,* 54 (1989), pp. 1383–1389.

Lieberman, M. (ed). 1981, 1983. Post-Harvest Physiology and Crop Preservation. NATO advanced study institutes series. Series A, Vol. 46. New York: Plenum Press.

McHenry, Robert (ed.). 1996. *Encyclopedia Britannica.* UK: Encyclopedia Britannica, Inc.

Meyer, L. H. 1960. *Food Chemistry.* New York: Reinhold Publishing Corporation.

Molins, R. A. 2001. *Food Irradiation: Principles and Applications.* New York: John Wiley & Sons.

Olaeta-Coscorroza, J. A. and P. Undurraga-Martinez. 1995. Estimacion del Indice de Madurez en Paltas. In: Harvest and Post-Harvest Technologies for

Fresh Fruits and Vegetables. *Proceedings of the International Conference, Guanajuato, Mexico, Feb. 20–24*, pp. 521–525.

Pantastico, E. B. 1975. *Postharvest physiology, handling, and utilization of tropical and subtropical fruits and vegetables.* Westport, CT: AVI Pub. Co.

Perry, R. H., D. W. Green and J. O. Maloney. 1984. *Perry's chemical engineer's handbook.* New York: McGraw-Hill.

Prosky, Dov and Maria Lodovica Gullino (eds.). 2009. *Post-harvest Pathology.* Springer.

Ramaswamy, H.S. and S. Ranganna. 1981. Thermal inactivation of peroxidase in relation to quality of frozen cauliflower (var. Indian Snowball). *Can. Inst. Food Sci. Technol. J.* 14(2):139.

Ramaswamy, H.S. and S. Ranganna. 1982. Maturity parameters for okra (*Hibiscus esculentus* (L) *Moench* var. *Pusa Sawani*). *Can. Inst. Food Sci. Technol. J.* 15(2):140.

Ramaswamy, H.S., S. Ranganna and V. S. Govindarajan. 1980. A nondestructive test for determination of optimum maturity of french (green) beans (*Phaseolus vulgaris*). *Journal of Food Quality* 3(1): 11–23.

Roberts, R. 1990. Postharvest biological control of gray mold of apple by *Cryptococcus laurentii*. *Phytopathology* 80(6): 526–530.

Robertson, G. L. 1992. *Food Packaging: Principles and Practice.* New York: Marcel Dekker Inc.

Ryall, A. L. and W. J. Lipton. 1979. *Handling, transportation and storage of fruit and vegetables. Vol 1.: Vegetables and melons.* 2nd edition. Westport, CT: AVI Publication Co.

Ryall, A. L. and W. T. Pentzer. 1979. *Handling, transportation and storage of fruit and vegetables. Vol 2.: Fruit and tree nuts.* 2nd edition Westport, Conn.: AVI Publication Co.

Salunkhe, D. K. and B. B. Desai. 1984. *Postharvest Biotechnology of Fruits.* Vol. 1 and 2. Boca Raton, Fla.: CRC Press.

Shewfelt, Robert L. and Bernhard Bruckner. 2000. *Fruit and Vegetable Quality: An Integrated View.* Boca Raton, FL: CRC Press.

Singh, R. P. and D. R. Heldman. 1993. *Introduction to food engineering.* Amsterdam: Academic Press.

Smock, R. M. 1979. Controlled atmosphere storage of fruits. *Hort. Review* 1:301.

Stevens, C., V. A. Khan, J. Y. Lu, C. L. Wilson, P. L. Pusey, M. K. Kabwe, E.C.K. Igwegbe, E. Chalutz, and S. Droby. 1998. The germicidal and hormetic effects of UV-C light on reducing brown rot disease and yeast microflora of peaches. *Crop Protection* 17(1): 75–84.

Swadling, I. R. and P. Jeffries. 1996. Isolation of microbial antagonists for biocontrol of grey mould disease of strawberries. *Biocontrol Science Technology*, 6 (1996), pp. 125–136.

Thompson, A. Keith. *Controlled atmosphere storage of fruits and vegetables.* 2010. Cabi; 2nd Revised edition.

Thompson, James F., C.F.H. Bishop and Patrick E. Brecht. Air Transport of Perishable Products. 2004, #21618. Regents of the University of California, Division of Agriculture and Natural Resources, Oakland, California 94608.

Thompson, James F., Gordon Mitchell, Tom R. Rumsey, Robert F. Kasmire, and Carlos H. Crisosto. Commercial Cooling of Fruits, Vegetables, and Flowers Rev. 2008. #21567. Regents of the University of California, Division of Agriculture and Natural Resources, Oakland, California 94608.

Thompson, Keith. 2003. Fruit and Vegetables: Harvesting, Handling and Storage, 2nd Edition. New York: Wiley-Blackwell.

US EPA. United States Environment Protection Agency. Food Irradiation. http://www.epa.gov/radiation/sources/food_irrad.html

USDA: United States Department of Agriculture. Food irradiation. http://fsrio.nal.usda.gov/food-processing-and-technology/food-irradiation.

Wikipedia: The Free Encyclopedia. FL: Wikimedia Foundation, Inc. http://www.wikipedia.org.

Willemot, C., M. Marcotte and L. Deschênes. 1996. Ionizing Radiation for Preservation of Fruits. In: *Processing Fruits: Science and Technology, Volume 1: Biology, Principles, and Applications*. L. S. Somogyi, H. S. Ramaswamy, Y. H. Hui (eds). Lancaster, PA: Technomic Publishing Co., Inc. Chap. 9, pp.221-260.

Wills, R. B. H. 1989. Postharvest: an introduction to the physiology and handling of fruit and Vegetables. Van Nostrand Reinhold, New York.

Wills, R. B. H., W. B. McGlasson, D. Graham, T. H. Lee, and E. G. Hall. 1989. Postharvest: an introduction to the physiology and handling of fruit and vegetables. Westport, CT: The AVI Publishing Co.

Wisniewski, M., C. Biles, and S. Droby. 1991. The use of the yeast *Pichia guilliermondii* as a biocontrol agent: Characterization of attachment to *Botrytis cinerea*. Biological Control of Postharvest Diseases of Fruits and Vegetables Wrkshp. Proc., Shepherdstown, W.Va., Sept. 1990. U.S. Dept. Agr.-Agr. Res. Serv. Publ. 92:160–176.

World Health Organisation. 1966. *The Technical Basis for Legislation on Irradiated Food.* WHO Technical Report Series, No. 316.

World Health Organisation. 1981. Report of the Working Party on Irradiation of Food. Joint Expert Committee on the Wholesomeness of Irradiated Food (JEFF). WHO Technical Report Series, No. 659.

World Health Organisation. 1994. Safety and nutritional adequacy of irradiated food. Geneva: WHO.

World Health Organisation. 1999. High-dose irradiation: Wholesomeness of food irradiated with doses above 10 kGy. Report of a FAO/IAEA/WHO Study Group. WHO Technical Report Series No. 890, Geneva.

Young, L., R. D. Reviere and A. B. Cole. 1988. Fresh red meats: a place to apply modified atmospheres. *Food Technology* 42 (9) pp. 65–69.

Index

Abrasion bruise, 215
Absolute humidity, 124
Accumulated heat units, 76, 77
Acetic acid, 50, 51, 64, 78, 290
Acetyl coenzyme A, 92
Active process (MA generation), 233
Aerobic respiration, 86, 101, 185, 200, 222, 231
Air circulation, 104, 146, 148, 167, 175, 176, 183, 210, 224, 303
Air exchange load, 177, 180
Air movement, 118, 175
Air velocity, 10, 119, 167, 174, 175, 176, 177
Alar, 64, 78, 103
Alternaria rot, 283, 284
Amyloplast, 40
Anaerobic respiration, 86, 93, 100, 209, 233, 239
Anthocyanins, 53, 54, 56
Anthoxanthins, 54, 56
Anthracnose, 284
Anti-nutritional factors, 57, 71
ATP, 25, 87, 90, 92

Bad cholesterol, 50
Berry, 13, 14, 16, 17
Biological control, 290, 291
Biot number, 136

Blue mold, 73, 284, 285, 286
Bracts, 15, 17
Brown rot, 282, 283, 291
Building transmission load calculations, 177, 178
Bulbs, 12, 17, 114, 279, 284
Bulk storage systems, 183
Bulky vegetative organs, 17

CA generation and control, 191
CA storage, 189
CA, maintaining, 193
CA, storage commodities, 191
CANDU® reactor, 250
Canopy shaker, 67
Carbohydrates, 9, 25, 38, 40, 41, 43, 45, 50, 51, 87, 91, 100, 240, 269
Carotenoids, 54
Carpels, 14, 16
Carrier irradiators, 252
Carrot, 12, 17, 18, 21, 52, 54, 65, 117, 129, 282, 293
Ceiling jets, 147
Cell wall, 23, 24, 25, 28, 29, 32, 33, 34, 40, 41, 43
Cell wall pressure, 33
Ceric-cerous, 272
Cesium-137, 241, 246, 247, 250, 251, 257, 274, 275, 278

Chilling injury, 8, 9, 21, 103, 302, 303, 304
Chilling sensitive commodities, 21, 168, 300
Chlorofluorocarbons (CFCs), 158, 159
Chlorophyll, 25, 40, 53, 55, 80, 286
Chlorophyll a, 53
Chlorophyll b, 53
Chloroplast, 25, 40
Cis fatty acids, 48, 49
Citric acid, 50, 92
Citric acid cycle, 50, 89, 90, 92
Clamshells, 228
Classification of fruits, 11, 12, 13, 14, 15, 16, 17, 18, 19, 20, 21, 22, 23, 302
Climacteric fruits, 19, 96, 99, 102, 103, 196
Cloudiness, 43, 44
Cobalt-60, 241, 246, 247, 250, 251, 255, 257, 270, 274, 278, 279, 280
Codex standard for irradiation, 245, 268, 274
Cold storage, 104, 132, 144, 145, 162, 167, 168, 174, 175, 176, 177, 180, 181, 182, 183, 184, 186, 188, 190, 192, 194, 196, 198, 202, 204, 206, 208, 219, 226, 231, 282, 283
Cold wall, 149, 150
Collenchyma, 28, 29, 31
Commercial forced air cooling, 149
Commercial irradiators, 280
Commercial shrink packaging, 230
Commercial vacuum cooling, 153, 154
Common fumigants, 299
Compression injury, 214, 215
Concentration gradient, 110, 197, 198, 209, 235, 236, 237, 238
Condensation, 110, 114, 115, 123, 130, 156, 173, 292, 304
Conduction, 132, 133, 134, 135, 136, 165, 178
Configurational isomers, 49
Consumer packages, 223
Container standardization, 212, 214
Container unitization, 214
Convection, 132, 133, 134, 135, 136, 154, 178, 194

Conveyor belt, 64
Cooling bay, 147, 149
Cooling coefficient, 138
Cooling effect, 108, 129, 189
Cooling methods, 132
Cooling rate curve, 139
Cooling rate, 131, 132, 137, 138, 139, 141, 143
Cooling, calculations, 129
Corn harvesting, 67
Corrugated fiberboard, 226, 228
Crispness, 24, 31, 32
Critical moisture loss, 109
Cryogenic cooling, 164
Cucumber, 11, 17, 18, 19, 20, 21, 39, 56, 65, 68, 77, 109, 119, 238, 283, 284, 304
Curie, 255
Curing, 114, 115, 116
Customs declaration form, 296, 297
Cuticle, 27, 28, 46, 109, 292
Cuticular membrane, 27, 109
Cytoplasm, 23, 25, 29, 33, 46

DC accelerators, 253, 254
Decimal reduction value, 267
Degree-days, 76, 77
Degree-hours, 76
Dehydrogenase, 90, 94
Delphi principle, 3
Dermal system, 26, 27
Dew point temperature, 123, 124, 175, 292
Dietary fiber, 41
Digenea simplex, 58
Diplodia rot, 283
Direct contact devices, 64
Diseases, 41, 150, 281, 282, 283, 290, 293, 296, 297
DNA, 25, 263, 264, 269
Dose rate, 258, 259, 260, 261, 272
Dosimeter, 272, 273
Dosimetry, 271, 272, 273
Drupe, 13, 16, 17
Dry bulb temperature, 120, 121, 122, 123
Dry fruits, 13

Durable crops, 4
Dynamic equilibrium, 33, 122, 123, 197

Edible film components, 240
Edible films, 239
Electromagnetic radiation, 136, 245, 246
Electromagnetic spectrum, 246, 248
Electron beams, 241, 270, 279
Electron transport system, 25, 90, 92
EMP pathway, 89, 91
Endoplasmic reticulum, 24, 26
Erwinia rot, 283
Essential fatty acids, 48, 269
Ethephon, 64, 78, 79, 103
Ethylene ripening, 197, 219, 298
Ethylene synthesis, 195
Evaporation, 107, 122, 123, 124, 136, 153, 157, 162, 189
Evaporative cooling, 157, 161, 162, 163, 186, 188, 189
Excitotoxicity, 57
External gas generators, 192
Extinction Point (EP), 101

Factors influencing quality, 75, 77, 79, 81, 83
Fats, 37, 38, 46, 47, 48, 49, 50, 85, 87, 91, 269
Fibrousness, 29, 31
Fick's law, 109, 110, 117, 198, 111, 200, 201, 203, 204, 208
Field containers, 146, 222
Field heat, 118, 127, 128
Filacell, 175, 185, 186, 187, 188
Firmness, 31, 73, 74, 81
Fixtures (for texture measurement), 36
Flavonoids, 54
Flavor, 38, 50, 71, 73, 74, 75, 86, 108, 210, 231
Flavor components, 56
Flavor compounds, 56
Floral, 15, 17, 18
Fog jet humidification, 185, 187
Food Guide, 37, 38
Food irradiation, 241, 242, 243, 244, 245, 248, 264, 25, 267, 268, 269, 271, 273, 274, 279

Food pipeline, 2, 6
Forms of radiation, 243, 245, 247
Fourier's law, 132
Freezing injury, 9, 72, 103, 104, 168, 304, 305
Fricke, 272
Fruit flies 295, 296, 301
Fruit thinning, 63, 78, 103
Fruit vegetables, 17, 18, 80
Fumigation, 219, 231, 293, 298, 299, 300
Fungicides, 9, 79, 239, 269, 290, 292, 302
Fusarium rot, 283

Gamma radiation, 244, 246, 250, 291
Gas diffusivity, 208
Gas refrigeration system, 159
Gluten, 46, 239, 240
Glycolysis, 25, 29, 89, 90, 91, 92, 93
Golgi complex, 24, 25
Good handling practices, 216
Gray, 258, 280, 286
Grey mold, 73, 282, 285, 290
Grittiness, 29, 31, 73
Ground system, 26, 28
Growth regulators, 9, 63, 64, 75, 78, 94, 102
Gynoecium, 14

Half cooling time, 138, 139, 140, 141, 143, 144
Hand harvesting, 59
Hand held harvesting devices, 61, 62, 63, 64, 65, 102, 187, 252, 275
Handling units, 213
Harvest, 1, 59
Harvest aids, 66, 70
Harvest containers, 213, 217
Harvest control, 63
Harvesting cages, 65
Harvesting stage, 79, 81
HDL, 50
Heat transfer coefficients, 136, 150, 178, 301
Hesperidium, 13, 15, 17
Hexokinase, 89
High altitude cooling, 163

Horticultural maturity, 80
Host defense mechanism, 286, 287
Humidifying pads, 188
Humidity ratio, 120, 121, 122
Hunter's (model), 72
Hydro cooling, 145, 150, 151, 164
Hydrostatic pressure, 24, 32, 33, 34, 39
Hypobaric system, 191, 205, 207, 208, 209, 210

Immature fruit, 18
Impact injury, 215
Infection process, 287
Insect control, 219, 264, 277, 295, 296, 297, 298, 299, 300, 301, 302
Insect infestation, 276, 278, 280, 295
Ionization, 248, 249
Ionizing radiation, 241, 244, 248, 249, 278
Irradiation application areas, 264
Irradiation dose, 267, 268, 275

Jacketed storage, 119, 175, 185, 186
Juiciness, 31, 32, 73

Krebs cycle 91, 92

Labeling of irradiated foods, 261
Latent heat, 107, 122, 123, 129, 130, 131, 154, 155, 158, 160
LDL, 50
Leafy succulent tissues, 17
Leafy tissues, 108
Legumes, 11, 14, 37, 39, 118
Lignification, 29
Lipid content, 47
Low pressure storage, 205, 206

MA generation, 233
MA maintenance, 234
MAP, 232, 233, 234, 235, 236, 237, 238
Mature fruit, 18, 286
Mean heat units, 76, 82
Mechanical harvesting, 61, 62, 63, 64, 65, 82, 222
Mechanical injuries, 27, 214, 215, 216
Mechanized harvesting, 60, 61, 213
Membrane storage, 202, 232

Membranes, 24, 33, 158, 197
Mesh bags, 227, 229
Mineral imbalances, 53
Mineral salt and water absorption, 108
Mineral salt distribution, 108
Minerals, 30, 31, 38, 40 51, 52, 74, 246, 250, 251, 269, 270
Miscellaneous heat loads, 178, 182
Mitochondria, 24, 25, 40, 90
Moisture loss, 27, 34, 107, 108, 109, 113, 114, 116, 117, 119, 150, 153, 175, 217, 218, 230, 231
Molded polystyrene, 228
Monosodium glutamate, 57
Morphological defects, 72

Neurotoxicity, 57
Newton's law, 134, 137, 138, 141, 142
Night cooling, 163
Nitrogen flush, 192
Non-chilling sensitive commodities, 168
Non-climacteric fruits, 99, 100, 102, 103
Non-ionizing radiation, 241
Nucleus, 23, 25, 40, 243, 245, 246, 247, 255, 263, 270, 288

Oils, 28, 37, 38, 46, 47, 48, 49
Omega 3, 48
Omega 6, 48
Onion harvester, 69
Optimum harvest maturity, 76
Organic acids, 25, 38, 40, 50, 51, 91, 100
Osmosis, 33
Osmotic pressure, 25, 32, 33, 34, 39
Outward diffusivity, 209
Oxalic acid, 50, 51, 58

Package icing, 145, 151, 152, 164, 219
Packages with vent holes, 219
Packaging requirements, 232
Pallet bins, 224, 225, 226, 228
Pallets, 147, 212, 214, 224, 225
Paper bags, 227
Paperboard, 211, 222, 223, 226, 276
Parenchyma, 28, 30, 31
Parkinsonian dementia, 57
Partial pressure, 110, 111, 113, 114, 123, 152, 161, 207

Particulate radiation, 243
Passive modification of the CA, 191
Passive process (MA generation), 233
Pathogen control, 275, 290
Pathogen entry, 287
Patulin, 285
Pectic substances, 24, 35, 40, 41, 43, 81, 91
Pectin methyl esterase, 44, 288
Pentose Phosphate Pathway 92, 93
PEPO, 13, 15, 17
Pepper harvesters, 68
Pericarp, 12, 15
Perishable, 2, 3, 4, 5, 129, 281
Permeability, 119, 167, 197, 198, 200, 201, 202, 203, 212, 221, 229, 232, 233, 234, 235, 236, 237, 238, 239
Pheophytin, 53
Phloem, 30, 31
Phospho-fructo-kinase, 90
Photons, 246, 247, 248
Physiological defects, 72
Picker pole, 62, 64
Pigments, 9, 25, 38, 40, 53, 54, 56, 240, 288
Pilferage, 219, 224
Plant pathology, 285, 286
Planting system, 63
Plasmaderma, 24
Plasmalemma, 23, 24, 25, 40
Plastic films, 212, 229
Plastic mesh bags, 229
Plastic netting, 224
Plastic-film bags, 231
Pome, 13, 14, 17, 27, 64, 78, 118, 282, 283, 286, 290
Postharvest chain, 5, 215
Postharvest losses, 1, 2, 3, 4, 5, 6, 7, 8, 9, 280, 282
Postharvest pathogens, 281
Postharvest quality defects, 53
Potato, 12, 17, 18, 20, 21, 37, 39, 41, 43, 47, 49, 52, 53, 56, 58, 65, 104, 109, 115, 133, 167, 190, 293, 304
Prebiotic, 42
Precooling, 10, 119, 127, 128, 177
Pre-harvest sprays, 293

Pressure cooling, 148
Primary dosimeter, 272
Probiotic, 42
Produce quality, 75, 94, 164, 168
Product load, 144, 177, 181
Protective tissues, 26, 27, 46, 119, 286
Proteins, 9, 24, 25, 37, 38, 46, 56, 85, 87, 91, 92, 239, 240, 269
Protoplasm, 40
Pruning, 63, 75, 78
Psychrometry, 120, 121, 123, 125
PUFA, 48
Pulpboard containers, 227

Quality perceptions, 74
Quarantine treatment, 242, 264, 298

RA storage, 184, 185, 187
Radappertization, 266
Radiant cooling, 163
Radiation (detection), 273
Radiation dose, 258, 259, 261, 263, 270
Radiation intensity, 255, 258, 261
Radicidation, 268
Radioactive decay, 247, 252, 255, 256
Radiochromic film, 273
Radura, 261
Radurization, 266
Recommended storage temperatures, 169, 170, 171, 172, 173
Refrigerant, 154, 155, 156, 157, 158, 159, 160, 161
Refrigeration requirements, 88, 130, 144, 145, 177, 179, 181
Refrigeration system, 130, 131, 146, 147, 153, 154, 155, 156, 157, 159, 160, 174, 184, 185, 186, 205, 206, 210
Relative humidity, 7, 9, 10, 34, 110, 111, 112, 113, 115, 118, 119, 120, 121, 122, 123, 146, 150, 151, 167, 168, 169, 170, 171, 172, 173, 174, 176, 184, 185, 186, 189, 196, 206, 209, 217, 239, 290, 292, 303
Respiration rate, 5, 8, 18, 88, 944, 95, 96, 97, 98, 100, 101, 104, 105, 106, 145, 182, 189, 198, 202, 204, 210
Respiratory Quotient, (RQ) 100, 101

RF accelerators, 254
Rhizopus rot, 282, 283
Rhodotron, 254
Rigid containers, 228
Rigid plastics, 228
Ripening of banana, 35, 36
RNA 25, 263, 269
Room cooling, 145, 146
Roots, 1, 12, 17, 27, 41, 56, 80, 108, 114, 115, 117

Saturated water vapor pressure, 110, 111, 112, 113, 114, 115, 121, 152
Sclerenchyma, 28, 29
Scrubber, 193, 194, 196
Scrubbers, 190, 193, 194, 196
Semipermeable, 24, 25, 32
Senescence, 5, 7, 18, 20, 80, 86, 96, 99, 189, 195, 196, 239, 262, 286, 303
Sensible heat, 127, 129, 177, 181
Serpentine cooling, 149, 150
Shaking, 61, 62, 66
Shipping containers, 214, 217, 222, 226, 227, 234
Shrink film wraps, 229
Sigmoid growth curve, 289
Sizes of fruits and vegetables, 21
Soft rot, 282, 284, 285
Solar collectors, 163
Sorosis, 13, 16, 17
Sour rot, 284
Specific gravity, 82
Standard pallets, 214, 215, 224
Starch, 25, 32, 35, 40, 42, 43, 44, 77, 84, 85, 91, 227, 229, 239
State point, 123, 124, 125
Stem, 12, 15, 17, 18, 28, 30, 37, 53, 282, 283, 284
Stem end rot, 282, 283, 284
Storage proteins, 46
Stretch film, 224
Structure (plant cell), 23, 24
Suction cooling, 147
Suction pressure, 34
Sugars, 25, 35, 40, 41, 42, 43, 51, 77, 81, 85, 87, 88
Synconium, 13, 16, 17

Tannins, 54, 56
Tartaric acid, 50, 51
TCA cycle, 50, 90, 91, 92
Temperature quotient of respiration (Q_{10}), 95
Texture profile analysis, 73, 74
Texture, 4, 5, 9, 27, 29, 31, 32, 33, 35, 36, 43, 45, 46, 71, 73, 74, 78, 81, 83, 108, 189, 279
Texturemeter, 73
Thermal conductivity, 132, 133, 178
Thermoelectric cooler, 165
Thermoelectric cooling, 165
Tomato, 11, 12, 18, 19, 20, 21, 44, 52, 54, 65, 76, 80, 109, 191, 215, 238, 283, 293, 304
Ton of refrigeration, 131
Tonoplast, 24, 25, 40
Toxic constituents, 58
Toxicological, 268, 274, 290
Trans fatty acids, 48, 49
Transit cooling, 164
Transpiration Coefficient, 116
Tree shaker, 66
Tree training, 63
Trellising posts, 70
Tristimulus colorimeter, 72
Tubers, 17, 27, 39, 41, 80, 114, 115, 118, 190, 279, 284
Turgidity, 24, 31, 32, 33, 34, 35, 39
Turgor pressure, 32, 33, 39
Typical curing room, 115
Typical psychrometric chart, 121

Units of radiation, 255
Universal Product Codes (UPCs)/bar codes, 221
UV, 159, 291

Vacuole, 24, 25, 28, 31, 32, 40, 50
Vacuum cooling, 145, 152, 153, 154, 162, 205
Valves, 28, 156, 194
Vant Hoff's Law, 95
Vapor absorption refrigeration, 159, 160
Vapor compression refrigeration, 157

Vascular system, 26, 30
Ventilated storage, 167, 183, 189
Vibration (and abrasion) bruise, 215
Vibratory devices (harvesters), 64
Vitamin A, 51, 54, 55, 58
Vitamin C, 51
Vitamins, 38, 40, 51, 52, 74, 240, 269

Water Vapor Pressure Deficit (WVPD), 111, 112, 116, 117, 118, 152, 292

Water vapor pressure gradient, 109, 110
Watery rot, 282, 285
Wax coating, 217, 226, 239
Waxes, 27, 234, 239, 240, 302
Wet bulb temperature, 122, 123, 124
Wooden containers, 216, 224
Wooden wire bound crates, 225
Wound healing, 116, 288

Xylem, 30, 31